TABLE OF CONTENTS

PREFACE

This book was required. As a former professor of military history at the American Military University (AMU) and a retired USAF Colonel, I wanted to teach a course on the effects of fighter aviation in war. In addition, I wanted to create a course that was a "hands-on" approach to fighter aviation history. Having flown USAF fighters for more than 20 years, I felt I had a good working knowledge of fighter aviation, but when I researched the subject I found I would have to ask my students to read scores of books to provide the background they would need. There was no "single-source" book that covered everything I wanted to cover. I determined to write that book. I wanted to write a straight-forward book in plain language that would not bore fighter pilots and at the same time that would be simple enough to be attractive to laymen as well as air power historians.

That is what I have done. I have covered some detailed thoughts about fighter flying in what Southern Americans might call "biscuits and gravy" language.

The overriding premise of the text is that the fighter[1] has been the key element in the air power equation and continues in that role today although this role may be changing with the advent of good, reliable, beyond visual range air-to-air missiles. This view has not been universally held over the years; however, it is a view that has been held by fighter pilots since the advent of the fighter. More and more historians are beginning to support this view.

Between World Wars, however, strategies based on bombardment aviation theories dominated almost all of the air forces of the world. In spite of WW I experience, the Air Power Doctrine that "the bomber would always get through"[2] pervaded the era between World Wars. The foremost disciples of Air Power in WW I, General Douhet (Italy), General Mitchell (U.S.) and Air Marshall Trenchard (Great Britain) were convinced the bomber would dominate in future wars. They had many senior adherents that dominated the air forces of the world. In this atmosphere, fighter pilots who reminded their bomber oriented leaders of the lessons of WW I were very roughly handled. Many fighter pilots risked their careers because they refused to stop purporting the effectiveness of the fighter in war. The disagreements were doctrinal, emotional and even fanatical at times. General Giulio Douhet wrote a book, Command of the Air, in 1921 that stated that the bomber would prevail over all other weapons. This view was adopted by most air forces the world over.

The U.S. Army Air Corps (USAAC) was not immune to this bias. General Billy Mitchell was court martialed for arguing the bomber case in the newspapers when his senior ground officers tried to smother his Air Power claims. The controversy caused him to resign his commission only to meet an early death in civilian life. However his followers who were still on active duty, adopted Mitchell's messianic zeal for a separate air force (like

the RAF in Britain). At the same time, this zeal was transferred to their promotion of the strategic bomber as the decisive weapon of war vice surface forces and fighters. The reaction to any challenge of this bomber promotion within the USAAC, no matter how logical, was fanatical opposition.

For example, Captain Claire Chennault, a brilliant and abrasive USAAC fighter theorist, chose to resign his commission in the U.S. Army Air Corps in the face of this emotionally supported bomber doctrine and its outright rejection of fighter doctrine based on WW I experience. To make matters worst, Chennault did not get on well with any authority that disagreed with him and he did not agree that the bomber was dominant. He had to go to China and accept a commission in the Chinese Air Force to prove his ideas on fighter defense and tactics in 1937. WW II proved that Chennault's ideas were correct even if his marketing skills were deficient. His performance with the American Volunteer Group ("Flying Tigers") in 1941 alone, has assured his place in air war history.

The Senior officers, who controlled the USAAC during the 1930s, were known to the fighter pilots as "the Bomber Mafia". This bomber bias was further evidenced by the fact that no head of the U.S. Air Forces was a fighter pilot until the 1980s when General Charles Gabriel became chief. Today with four straight fighter pilots named as Air Force Chief Of Staff, we have come full circle and there is talk today of "A Fighter Pilot Mafia".

In Nazi Germany, Adolph Galland, Werner Moelders, etc., like Chennault, were hamstrung by Luftwaffe superiors, who did not understand fighters and how they fit into the air war equation. This blind adherrence to the theories of Douhet by senior Luftwaffe leaders may have cost Germany the war. This failure was exacerbated by another decision based on Spanish Civil War experience to stick with medium bombers for European strategic targets and slow dive bombers for the close support of troops. The Germans never developed a successful heavy bomber or successful long-ranged fighter escort tactics for their strategic bombers.

On the other hand, those who understood fighters and their roles were extremely successful particularly since the beginning of WW II. Fighters and air superiority were crucial in WW II and air supremacy has likewise been foremost in every Israeli success against the Arabs for more than 50 years. This predominance of the fighter continues. The devastating use of fighters by Allied Leaders in Desert Storm was our most recent example.

Our newest, most capable and most expensive fighter, the Lockheed F-22, is in development now, and is scheduled to deploy right after the turn of the century. The lessons of military history are clear. If we are to prevail in tomorrow's world, we need the F-22 armed with state-of-the-art air to air missiles (AAMs), as quickly as available, to help prevent us from losing the next war. For those who insist it is too expensive, I would commend the words of General W.T. Sherman who said; "War is hell". Sherman was on the winning side and if winning in war can be defined as hell, then losing in "hell" must be the most expensive thing that can happen to a nation.

Nor are the best machines enough, as an essential ingredient of every fighter weapons system has been the pilots that fly them and the men who maintain them. This human element of the system has proven as essential as the machine itself.

Mr. Gil Robb Wilson, World War I aviator, an ordained Methodist Minister, and an editor and writer for "Flying" magazine for years, best captured the total essence of fighter flying in the last verse of his poem "Briefing":

"With God in your guts,
Good men at your back,
Wings that stay on
And --- Tallyho!"[3]

He also quite inadvertently provided the title for this book: ***Wings That Stay On.***

1.. The term "fighter", as I use it throughout the text, applies to the fighter weapons system in its entirety which includes the airframe, the pilot (crew) and all of its armament. In WW I, many countries used the term "pursuit" vice fighter. The U.S. Navy began to use fighter in the 1920s and the universal use of fighter did not appear until WW II. The USAF adopted fighter officially at the end of WW II when they changed the "P" designation to "F"; e.g. "P-51" became "F-51".

2. . Quote by the Prime Minister of Britain Stanley Baldwin in 1932. See: John Terraine, *A Time For Courage The RAF In The European War 1939-1945*, MacMillan Publishing Company, New York, 1985, p. 13.

3. . Gill Robb Wilson, Airman's World: "The Briefing".

CHAPTER 1

WORLD WAR I

INTRODUCTION

The premise of this book is that fighter aviation is one of the most important components in air warfare. In fact, we will contend and prove that fighters have been the key in achieving control of the air (air superiority and air supremacy) in almost every past war. Without control of the air, it is extremely difficult to win a conventional war. Few of the original adherents of air power seem to have realized the dominant role of the fighter.[1]

NOTE: In this book, for clarity, we will use the term "fighter" as it is currently used by the U.S. Air Force. Fighter will be used even during discussion of WW I aircraft although WW I terminology was usually "chasse", "pursuit", or "scout". Actually the term might be more appropriately "tactical fighter" as it encompasses all of the possible missions of fighter aircraft both air-to-air and air-to-ground, to include: air superiority and supremacy, interdiction, and close support.

The early prophets of air power (Douhet, Trenchard and Mitchell) were entranced with the pervasiveness of the bomber. Very little attention was given by these early air power theorists to the fighter. They were captivated by the prospects of concentrating airborne firepower on an enemy's total war-making capabilities using the initiative provided by use of the Principle of the Offensive.[2] In this scenario, defensive air forces would be put at a marked disadvantage as they had no way of knowing where the attacking air forces would strike and could not concentrate fast enough to meet a rapid, well-directed air offensive. These men felt that the difficulties of detecting, intercepting and destroying in-coming bomber aircraft were just too great.[3] The payload and range requirements for these strike aircraft dictated that they had to be large, bomber-type aircraft. As late as 1939, experts still believed that the bombers must always get through. WW I had taught us otherwise,[4] but memories were short and it remained for WW II to dampen this mythology. In the first two years of WW II, the British found that, without control of the air, their unescorted (not accompanied by fighters) bombers sustained prohibitive losses.[5] The USAAF discovered this fact on its own with its four engined bombers over Germany in 1942-43.[6]

The radar, command and control systems, and operations centers of WW II, first demonstrated during the Battle of Britain, proved the fallacy of the bomber will always get through ideas. Bombers were unable to satisfactorily operate when the enemy had gained air superiority that was provided by swarms of modern fighter aircraft.[7] Douhet had it right when he outlined the proper air objective as control of the air, but he and Mitchell and Trenchard erred when they concluded this control could be gained by unescorted bomber aircraft alone while using conventional high explosives. It was the "Pursuit" (fighter) aircraft that determined control of the air in WW I, and fighters again dominated in Spain and Finland, between wars. Yet the "bomber mafia" marched into WW II still believing their original doctrine that bombers alone would prevail in gaining control of the air. ME-109s, Spitfires, Hurricanes, and Zeroes abruptly contradicted these zealots. Until the advent of nuclear weapons, air operations during WW II to gain control of the air were restricted by the radius of action of the attacker's *fighter* aircraft.[8] In conventional war, and with deference to modern missile defenses, this is still true today.

Today, development of new radars, electronic warfare, stealth technology, air-to-air and ground-to-air missiles, are changing fighter weapons systems dramatically. The race goes on.[9] It is the purpose of this book to describe that race from start to finish and determine what it all means.

THE BEGINNINGS OF FLIGHT

Before man could use air forces in war, he first had to fly. If we are to believe history and our myths, man's attempts to fly began during the days of antiquity. Greek mythology places the first aeronauts as Daedalus and Icarus, but Icarus's fate from flying too close to the sun did nothing for the credibility of the Greeks.[10] Many other conceptions of how to achieve manned flight are recorded during the middle ages. These include a number of unsuccessful tests and attempts. The most famous concepts are probably those of Leonardo da Vinci who left over 500 sketches and 35,000 words on the possibilities of flight.[11]

Two centuries before, Roger Bacon in England had prophesied that men would fly someday, but his idea, an ornithopter, is only remotely related to the helicopter of today. However, the airscrew discussed by Bacon was correctly forecast as valuable to flight by the facile Leonardo. The Chinese used kites for thousands of years and perhaps as long ago as 400-500 B.C. The first wartime airborne reconnaissance was probably flown by one of these kites about 200 B.C. by the Chinese General Han Hsin when he measured the distance he needed to burrow to get under an enemy city's walls using the kite's tether as a distance-measuring stick.[12]

Beginning in the late-18th century, man's desire to fly began to take fruition. First the balloonists, then kites and gliders became of scientific interest. The French revolutionists were the first to take a military interest in balloons. The capability of observation balloons to provide useful military intelligence was proved to the new French government by the noted chemist Jean Marie-Joseph Coutelle in 1794. On April 2, 1794, the French formed the world's first Air Force and the Companie d'Aerostiers was assigned to the command of the newly commissioned Captain Coutelle.

These aerostiers were later disbanded by Napoleon himself after his return from Egypt even though they had proven their capabilities handily at the Battle of Fleurus. Balloonists were then relegated to relative obscurity for 50 years by the conservative military. They next appeared on the Union side at First Bull Run in 1861 during the American Civil War.[14]

Consolidated under the direction of Professor Thaddeus Lowe, balloons played an integral part in Civil War battles on the Union side, particularly in its Eastern campaigns. On the other hand, the South flew only one balloon during the war. According to Southern General James "Ole Pete"Longstreet, this balloon was a real tragedy as it was sewn from a number of ladies silk dresses. As Longstreet saw it, the tragedy was in the disappearance of these gowns from southern soirees. The southern ladies were dressed much plainer at their dances and parties.[15]

Balloons then participated in every war thereafter until the end of WW II. For example, during WW I, the French manufactured some 4,000 observation balloons and the Germans nearly 2,000. When trench warfare began, casual observers could locate the front from a distance using the French "Caquots" and German "Drachen" balloons strung behind the trenches.[16]

As the balloons promoted manned flight, the scientists pressed on with experiments on flying machines that were heavier-than-air. One of the foremost was George Cayley, who riding the groundswell of scientific inquiry in 19th century England, spent his life studying "flying machines". His drawings, models, and experiments were full of modern ideas (as well as some worthless concepts). Cayley's student Samuel Henson proposed an "aerial carriage" that caught the imagination of the public, but that went relatively unsupported by the scientific community. However, Henson's work inspired Cayley and in retrospect, many of his comments on Henson's work were truly pioneer efforts. Cayley actually achieved gliding flight combining Henson's con-

cepts with his own. Cayley died in 1857.[17]

Cayley's work was followed by a frenzy of scientists and enthusiasts who built gliders particularly in France. Le Bris, Du Temple, and Penaud, each contributed to the growing data on aeronautics. The British again emerged with Francis Wenham who built the first wind tunnel for testing. Ader in France, Langley and Chanute in the U.S., and Otto Lillenthal in Germany followed by developing concepts, data, and experiments. Lillenthal, in particular, used a scientific approach and developed equations for lift and methods of research for control of his machines after they were airborne.[18]

But it remained for two mechanically-inclined Americans to put all of this activity together in a logical, systemized fashion and to thoroughly document each step. Orville and Wilbur Wright were sons of a Methodist Bishop in Dayton, Ohio. Neither had attended college, but both were clever and skilled with mechanical systems and had developed modern printing and bicycle systems which had proven both practical and profitable. During the 1890s they caught the "flying bug" and dedicated many long hours of study to the subject. Their practical minds researched the past for everything written on flying and they then tested the premises with their own models and experiments. They found the work of Cayley, Penaud, Lillenthal, and Chanute extremely helpful (Langley was looked on as a competitor). Through experiments with gliders and later powered models, they developed their own roll control mechanism, found Lillenthal's lift equations to be in error, applied their own newly developed internal combustion engine to the glider, and starting over from scratch, developed a proper aircraft propeller. When they ran into problems with sideslip, they developed and properly emplaced a rudder system that permitted controlled turns. In December of 1903, the Wrights first flew a heavier-than-air machine that left the ground under its own power and which was able to steadily duplicate its initial controlled flight. The Air Age was born.[19]

AIR WAR

It was a very short 10 years and eight months from the first flight at Kittyhawk and the outbreak of WW I. The Italian Army first used aircraft in war against the Turks in 1911. They dropped small bomblets and grenades on troops, and performed observation, reconnaissance, and liaison missions.[20] Incidentally, it was this North African campaign that first inspired the great air power theorist, Giulio Douhet.[21] By 1912, the French Army had more than 250 aircraft, and with war clouds gathering, all of the powers of Europe began to build air arms.[22]

THE MISSION

As the war began, it became apparent that the machine gun and barbed wire had limited the effectiveness of cavalry for reconnaissance and airborne observation had become very important to the ground forces. Aircraft could sail serenely over the defenses and see and photograph the terrain and defenses below. Balloons and aircraft were essential to direct artillery fire. All of the armies began the war with unarmed aircraft. Many had official policies that forbade their aircrews to carry arms. But in a war, policies change and it became clear to officials and aircrews alike that the only way to stop harmful observations and bombings was to arm the aircraft. As the more aggressive pilots armed, it became imperative to the others that they had to arm in self-defense, particularly after they had been attacked.[23]

Below: *A Vickers FB5 "Gunship" powered by a 100 hp Gnome Monosowpape engine. One of the first "Pushers". Note the high drag superstructure needed to clear the pusher propeller and the field of fire for the gun forward.*

As WW I progressed , air forces on both sides had a new problem. Their air forces, including both airplanes and balloons, were initially designed solely for observation of ground forces, but experience quickly pointed out that to prevent enemy observation and bombing of friendly forces was equally important. The real task was to achieve air superiority, or air supremacy,[24] so you could perform any air mission and your enemy could not. It was this mission in WW I that was the genesis of the fighter.[25]

The French, who had avidly adopt-ed aviation before the war had the widest range of flying machines. By the end of August 1914 when the war began, France had 29 squadrons. The British, who started to build up in 1912, initially had to use French equipment. The Germans had concentrated on Zeppelins before the war for bombing missions. Germany had a fleet of seven Zeppelins at the start of the war along with six smaller airships. German airplanes were specifically designed for observation purposes. In fact, initially, some forward-thinking authorities imposed speed limits on German aircraft to ensure they flew slow enough for adequate observation![26] Germany had about 240 airplanes of all types as the war opened.

AIRCRAFT ENGINES

The major problem Orville and Wilbur Wright had in building their flying machine was finding and developing an engine light enough and still powerful enough to reliably provide adequate power for the machine. An engine that was light, powerful, and reliable was a tall order in 1903. For a fighter, this requirement became even more stringent as its job was to outperform other aircraft. It has been the dicta of fighter analysis since its inception that a fighter is only as good as its engine. Obviously, the engine must perform reliably and run. In the end, there is little difference in an operational loss where an engine fails and a combat loss where enemy action is the cause, particularly if you are flying over enemy lines. Also, it is the ratio of thrust (produced through engine power and the prop in prop aircraft) to both weight and drag that is crucial to performance (i.e., speed, acceleration, rate of climb).

4

Therefore, the lighter and more powerful an engine can be and still be reliable, the better. However, since it is the ratio of thrust to weight and drag, the lighter the airframe (but still sufficiently strong) the better, and the lower the drag through efficient design, the better.

During WW I, there were two basic types of aircraft engines. The first was the rotary as personified by the French Le Rhone. (The Germans used a copy of the Le Rhone called the Oberursel.) The rotary engine featured a stationary crankshaft with its air-cooled, radially arranged cylinders revolving around the crankshaft with the prop rotating with the cylinders. Its advantages were an outstanding power-to-weight ratio, it ran smoothly (the rotating cylinders acted like a flywheel and dampened vibrations) and it was easily air-cooled (eliminating the need for heavy cooling systems).[27]

Its disadvantages were the excessive torque generated by the engine mass as it rotated with the prop, and the fact that this rotation was self-limiting as far as increasing power was concerned. To increase the speed of revolution of the prop (measured in revolutions per minute or RPM) on a rotary engine produced a loss in the form of torque of 20% or more of the additional energy developed. This imposed an upper limit of about 200 HP on practical rotary designs. As it was, this rotation of the entire engine mass resulted in many failures of cylinders, crankshafts, etc. For example, the Gnome Monosoupape rotary used in the Nieuport 28 was so unreliable it was despised by the American pilots that flew it, although most really liked the airplane. The higher power-to-weight ratios paid off in overall aircraft performance when the engines performed as advertised.[28] The Nieuport was extremely fast, as well as nimble. Unfortunately, it had a habit of losing the fabric on its upper wing at high speed.[29]

The rotaries also had two other disadvantages: 1) They had a higher specific fuel consumption (ratio of about 7-5 with fixed engines) and 2) Without independent lubrication systems they were forced to mix lubricant with the fuel (like modern two-cycle engines). The only lubricant that performed well in aircraft engines at altitude when mixed with gasoline was castor oil.[30] In wartime, where disease and diarrhea are often rampant, one can only imagine the effects on the pilots from constantly being exposed to castor oil fumes and smoke in those open cockpits.

The Germans preferred the heavier water cooled in-line engines made by Mercedes, Daimler, and BMW. These engines were developed from automobile racing engines, and being heavier, required heavier aircraft structure to support them in the airframes. But as power requirements exceeded 200 HP, these stationary engines were required. The German engines were quite reliable in most cases, but most experts today rate the Allied V-8 Hispano-Suiza used in both the SPAD and SE-5A as perhaps the best engine of the war. The V-12 American Liberty was too late to really affect the war. Each year of the war the horsepower requirements increased. Engines started at around 100 HP in 1914. Four hundred horsepower was the minimum projected requirement by the end of 1918. If the war had continued one more year, the Liberty, with its 400 HP, would probably have been much more popular.[31]

Finally, as the war progressed, engine designers benefitted from improvements in metallurgy. Lighter, stronger metals were

developed and under the spur of wartime requirements, engineers were able to design engines of greater power that were lighter overall than the earlier designs. These improvements in engine technology particularly benefitted the stationary designs. This development continued after the war. One can grasp how much progress was made between wars when one realizes that the WW I Liberty engine of about 400 HP at sea level, had about the same displacement as the Rolls Royce Merlin of WW II and the Merlin developed approximately 1650 HP from sea level all the way to 25,000 feet!

The rotary engine with its great power-to-weight encouraged designers to design light, nimble, maneuverable fighters with high rates of climb, ability to roll, and high rates of turn. Good examples of this group were the Bristol Fighter, the Moranes, the Fokker Eindecker, the Sopwith Camel, the Nieuport, Fokker Tri-plane, and the Sopwith Triplane. A skilled pilot could utilize the high torque of the engine to

maneuver extraordinarily. On the other hand, this torque could be exceedingly dangerous for inexperienced pilots particularly when landing or making turns close to the ground. These aircraft were usually slower and not as stable as the heavier, stronger, fixed-engined fighters. Nor could most of the rotaries dive with these aircraft. In fact, high speed dives with some of these lighter structured aircraft could be dangerous.[32]

The fixed engine aircraft were generally more stable, faster, stronger, and able to sustain high speed dives. Later models like the SE-5A, SPAD, and Fokker D-VII also were very maneuverable, but they usually depended on superior speed and energy conversion to maneuver and initiate and terminate fights. One cannot fight what one cannot catch. Fifty years later Colonel John Boyd coined this tactic as "Energy Maneuverability".[33]

AIRCRAFT ARMAMENT

None of the aircraft that opened the war on either side were specifically designed for fighting. In some cases,

it was left up to the aircrews to devise their own weapons. During the opening days of the war, practically everything was tried, including rifles, shotguns, dropping grenades, darts or flechettes, and grapnels extended by a cable below the plane.[34]

The Allies even tried blunderbusses with smooth bores using grape shot and chain-linked shot. It soon was apparent to most that the machine gun was the weapon of choice.[35]

Just as designers had problems finding powerful, light aircraft engines, the ordnance experts had great difficulties finding and developing suitable machine guns. The problem with the guns in 1914 was that most were too heavy and the frail aircraft of the day lost signifi-

Below: *The SPAD VII powered by a fixed 150 hp Hispano Suize V8. The original "mount" of the French Ace Georges Guynemer. Armament 1-7.7mm. Vickers Mg.*

cant performance when the machine guns were installed.[36]

Interestingly, all of the solutions developed in England, and Germany had their roots with American gun designers or inventors. The Vickers, Spandau, and Parabellum, were all developments of the Maxim machine gun designed by Hiram Maxim, a U. S. inventor. The lightest machine gun of the day was the Lewis which was air-cooled and drum-fed and weighed about 25 pounds. The Lewis gun was designed by Colonel Isaac Newton Lewis an American Army officer. The Maxim was a water-cooled model and was much heavier. Both designs were offered to the U.S. Army and turned down prior to WW I. Both men went to Europe and sold their designs.

To adapt Maxim based models, it was required to devise an air-cooled version which of course was much lighter. The first adaptation was the German Parabellum which was air-cooled and drum fed. This gun closed a void for the Germans who had no aircraft machine gun model available to counter the Lewis equipped British planes early in the war. The Spandau was the air cooled, belt-fed version of the Maxim that was employed by the Germans after the drum-fed Parabellum. The belt-fed Spandau and British Vickers (both copies of the Maxim adapted for airborne use) allowed hundreds of rounds to be fired without reloading. This was a major improvement when compared with the magazine-fed guns with about 50 rounds per magazine.

The French used a version of the Hotchkiss machine gun, their stan-dard infantry machine gun. This gun was also used by the American Army prior to the war. Lewis had been turned down on the Lewis gun by American ordnance who preferred the Hotchkiss to the lighter Lewis. The Lewis actually performed better on aircraft than it did in the mud and grit of the trenches. The Hotchkiss was heavy and difficult to operate while airborne. In mid-1915, the French changed to the Lewis.[37] When the British switched to the Vickers as their primary aircraft gun, the French also adopted it.[38]

The use of machine guns posed several new problems for these embryo fighter pilots. First of all, there was a problem with weight. Machine guns were heavy and so was the ammunition. Early aircraft such as the British Avro 504 and the French Farmans could not carry even the lightest of the available machine guns without serious degradation in their performance. The first attempts at interception failed as the aircraft could not reach sufficient altitude to attack the observation ships.[39] This problem resulted in many pilots being told to forget the machine gun idea.

Another problem involved air-to-air gunnery. Vintage aircraft were a maze of struts, control surfaces, and wires. This structure on your own aircraft would often be brought into the gunner's line of fire as he tried to track the enemy aircraft. Firing guns at another maneuvering aircraft was nearly as dangerous to the firer as to the target. More than one aircraft staggered home on a wing and a wire damaged by his own friendly fire. Some did not come home. One can imagine how really "hairy" this could be as observers tried to track targets moving in three dimensions from a platform moving the same way. Pilots soon hit on aligning the guns with the longitudinal axis of the aircraft to cancel out these problems. They could then aim their guns by aiming the aircraft and the enemy's evasive maneuvers could be canceled out by matching his movements with those of their own ship. In the case of the pusher fighters with this fixed arrangement, they no longer had to worry about hitting their own ship and gunnery was much simpler. For the tractor pilots, how to fire through the propeller remained a problem.

But even with a good field of fire, air-to-air gunnery was new and complicated. Methods of sighting had to be developed. Equipment was needed to estimate range and "angle-off" (where to point the aircraft nose[gun] to hit a turning, maneuvering, target aircraft); not to mention a system to align the fixed guns accurately with the axis of the aircraft. The movable guns had to be mounted so that free movement and rapid tracking was possible. The physics of firing spinning projectiles laterally in 100 MPH airflow had to be studied. Many of these problems were not satisfactorily solved until WW II or later. Pilots in WW I solved most of this complexity by flying their aircraft up so close they literally could not miss their targets. During WW II, many aces did the same thing. Many never learned to be marksmen.

By 1918, all of the frontline fighters were using at least two machine guns. Interestingly, many frontline fighters (except in Russia, Germany, and Britain) were still comparatively armed in 1939. This rapidly changed with the advent of WW II.

Faced with the fact that some of the best performing and stable aircraft of the time were single-seat (less weight) and tractor powered (prop mounted forward), WW I fighter pilots were faced with a major problem: The propeller was in the way of forward-firing guns aligned with the fuselage of these aircraft. Obviously, some type of system to fire through the propeller was necessary for tractor fighters.

There were any number of compromises devised. New and better-performing pusher aircraft were designed, particularly by the British. This allowed a full field of fire forward. First came the Vickers "Gunbus" in 1914, a two-seater aircraft with the observer in front of the pilot and a free swiveling machine gun used by the observer.[40] Some pilots also installed a second fixed gun aligned with the longitudinal axis of the aircraft. This was Britain's first fighter and with its forward-firing armament had the advantage in the early air-fighting. We will see other pusher designs that gained advantage in 1916. The FE-2B was a two seater similar in configuration to the older "Gunbus" while the DH-2 was a single-seater, pusher with fixed armament. Both were instrumental in ending the "Fokker Scourge".[41]

These viable pusher fighters armed with the light Lewis machine gun gave the Allies the early edge in air superiority. It took a while for the Germans to develop an aircraft machine gun. Until the Parabellum machine gun was developed in November 1914, the Germans fought with rifles, carbines, and shotguns versus the early pushers armed with Lewis machine guns.[42]

Several French and British tractor models installed guns to fire outside the propeller arc, either on the upper wing of biplane fighters or to fire at an angle to the side of the longitudinal axis. The upper wing mount had the disadvantage of the aircrew having to stand to reload the gun.[43] This bizarre arrangement led to some exciting moments for the pilots in single seat aircraft. Consider the case of Lt. Louis Strange. Strange, who in August of 1914 was credited with arming the first aircraft of the war with a machine gun, quickly discovered the Farman aircraft he was flying could not climb above 3,000 feet altitude while the German observation plane he was attacking was at 5,000 feet. Strange had then been reprimanded and told to forget air combat. But in April of 1915, we find him in a single-seat Martinsyde fighter armed with a upper wing Lewis machine gun. Strange attacked, but his Aviatik adversary was too high and he could not gain an advantage. He emptied his 47 round drum at the German observation plane and had to reload. This involved un-strapping from his seat, standing up, holding the flight controls between his knees and replacing the Lewis gun magazine. This particular drum was very stubborn and refused to budge. As Strange tugged on the drum, the aircraft stalled, flipped inverted, and entered a spin with Strange still clinging to the magazine. Fortunately, the magazine still refused to budge. Strange was able to climb back up into the cockpit (he was still inverted) and recover the aircraft to level flight. He hit the bottom of his wicker seat on recovery with such force that pieces of his seat jammed his controls. On landing, he had to explain to his CO how he had managed to destroy the cockpit of the Martinsyde. Unbelievably, Strange survived the war.[44]

Captain Lanoe Hawker was another hero of this period. He rigged a single-shot hunting rifle to fire outside the arc of his propeller on the port side of his Bristol fighter. He later replaced the rifle with a Lewis machine gun. An excellent pilot and fearless, his motto was "Attack Everything" and he perfected a method of crabbing (yawing his aircraft sideways) his aircraft to bring his guns to bear. For destroying three machine-gun armed observation planes in the Spring of 1915 while using his single shot rifle, Hawker won the Victoria Cross.[45]

GUN SYNCHRONIZING GEAR

As we have seen, the participants in WW I desperately needed some method of firing machine guns through the propeller. As incredible as this may seem, the technology existed and had been developed by someone in three or four nations prior to, or just following, the outbreak of the war. In 1913, a Russian designer, Lieutenant Poplavko developed an interrupter gear that was later fitted to the Sikorsky S-16 aircraft. In the same year, Franz Schneider, the chief designer of the German company L.V.G., patented a synchronizer to fire through the airscrew of tractor aircraft. The synchronizer was installed in an L.V.G. design, but the aircraft was destroyed in a crash and no further interest was shown in the equipment. The Edwards brothers in England proposed plans for a similar device in 1914 only to be pigeon holed by the British War Office.[46] In

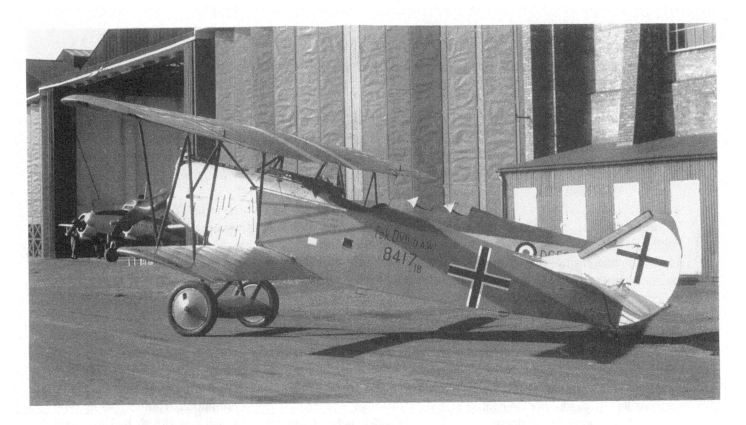

France, Raymond Saulier was working for the Morane company on the problem. By June of 1914, he had a device mechanically linking the trigger of a Hotchkiss machine gun to an aircraft engine's cam. Unfortunately, the Hotchkiss would sometimes overheat and rounds tended to "cook-off" on a delayed basis which endangered the propeller. Saulnier was undaunted and installed metal deflectors on the propeller to take care of these "stray" rounds.[47] The French government, like the Russian, British, and German, showed little interest.

In the Spring of 1915, Roland Garros, a pre-war stunt pilot went to the Morane Saulnier Company to help him resolve his difficulties in achieving air-to-air victories. He talked Saulnier into equipping a Morane Saulnier monoplane with the deflector blades without the additional weight and complexity of the interrupter gear. The result was immediate success. Garros shot down three German observa-

tion planes in a few days. He was the terror of the German skies for several weeks. Unfortunately, after a prolonged gliding attack over Courtrai, his Le Rhone engine refused to restart and he was forced to land behind German lines and was captured before he could destroy his aircraft. The Germans recognized Garros and noted the deflectors on his prop and realized their significance.[48]

The Germans summoned the young Dutch aeronautical engineer Anthony Fokker (whose genius had been spurned by the Allies) to view the deflectors on Garros' captured machine. Fokker did not think much of the design. He stated the deflectors were as big a threat to the attacker as the target, due to prop vibration and possible ricochets. He proposed a synchronizer system, designed and built a practical system in three days and went to Berlin to test fire the system using a Fokker Eindecker (monoplane) very similar to the Morane Saulnier Parasol that Garros flew. To his surprise, the Ger-

man authorities were very skeptical of his design because it did not use deflectors. Even after significantly successful ground firing tests and air-to-ground strafing demonstrations, the conservative Germans insisted on a combat test. Only after putting on a German uniform and actually getting airborne on a combat sortie, did Fokker rebel against shooting down an Allied plane just to test his invention. He returned to the German base, landed, and said "to hell with it". He then told the Germans he was through testing and to do with the design as they saw fit. Fokker's design put the Germans a year ahead in aircraft armament and was to cost many Allied lives.[49] The "Fokker Scourge" had Begun.

The Allies finally got effective gun synchronizers in late 1916 almost

9

a year after Fokker's system was operational. However, the Allied synchronizer equipped aircraft did not appear in any great numbers until the early summer of 1917.[50]

CHANGING FIGHTER DOCTRINE

Unfortunately for the Allies, the Fokker Eindecker with a forward-firing synchronized Parabellum machine gun occurred just before the maturation of fighter doctrine as conceived by a young German fighter pilot, Oswald Boelcke and another stalwart, Max Immelmann. These ideas matured during the battles fought around Verdun in 1916. The combination of a superior weapon system and coherent fighter doctrine was devastating for the rest of the war for the Allies. During the last part of 1915 until mid-summer 1916, Allied losses soared. This period was dramatically known as the "Fokker Scourge". Boelcke's ideas as outlined in his "Dicta" were still being used by fighter pilots on both sides 25 years later during WW II:

Boelcke's Dicta

1.Always try to secure an advantageous position before attacking. Climb before and during the approach in order to surprise the enemy from above, and dive on him swiftly from the rear when the moment to attack is at hand.

2.Try to place yourself between the sun and the enemy. This puts the glare of the sun in the enemy's eyes and makes it difficult to see you and impossible to shoot with any accuracy.

3.Do not fire the machine guns until the enemy is within range and you have him squarely within your sights.

4.Attack when the enemy least expects it or when he is preoccupied with other duties such as observation, photography, or bombing.

5.Never turn your back and try to run away from an enemy fighter. If you are surprised by an attack on your tail, turn and face the enemy with your guns.

6.Keep your eye on the enemy and do not let him deceive you with tricks. If your opponent appears damaged, follow him down until he crashes to be sure he is not faking.

7.Foolish acts of bravery only bring death. The Jasta must fight as a unit with close teamwork between all pilots. The signal of its leaders must be obeyed.

Except for #6 these are still good guidelines for air-to-air combat.[52] Number 6 only applies when there is no danger of exposure, i.e., no one else is around (when can you be sure?) or your tail is protected by wingmen.

The term "Fokker Scourge", in retrospect, probably was not justified as there were very few Eindeckers (never more than 150) and the aircraft designed as fighters were not that inferior to the Fokker monoplanes.[53] But if the Eindeckers were not that superior as aircraft, their gun system was, as belt-fed Spandaus (which replaced the earlier Parabellums) allowed 500 continuous rounds while the French and British had to reload frequently (every 47 rounds for the British equipped with Lewis guns). Organized by Boelcke into Jastas (squadrons) and flying and fighting as squadrons, according to estab-

lished doctrine, the Germans markedly improved their fighting efficiency. Losses were very high among the British Royal Flying Corps. Most of the losses were from units flying the British Experimental observation and bombing aircraft (BE-2Cs) which had been produced in large numbers even though they were known to be obsolete. Large numbers of new pilots were lost before they could develop into veterans. In the British Parliament, these young men, the cream of British youth, were called "Fokker Fodder".[54]

During the Battle of Verdun, the Eindeckers went all-out to gain air supremacy over the battlefield, prevent French reconnaissance from the air, and to open the way for German close support aircraft to strafe the French troops in the trenches in front of the German attack. Initially they were very effective. The French realized they had to deny the

airspace to the Germans and stop these attacks. They moved two groups of their best fighters into the area and a battle raged for control of the air. The French Nieuports assigned included the famous French "Storks" with several famous aces such as, Charles Nungesser, Jean Navarre and Georges Guynemer. The overall commander was Tricornot de Rose, Air Chief of the French 2nd Army. Among his units was the Lafayette Escadrille made up of Americans flying for France and led by the redoubtable ace, Raoul Lufberry.[55]

The battle raged from February to June 1916. During this time, Boelcke, who was the defacto air commander over Verdun for the Germans, refined his ideas for the German Air Service. Never again would the Germans fight individually.[56]

The French under Petain held Verdun on the ground, and they flooded the air with their Nieuports. Even though the Germans destroyed nearly 100 French aircraft while losing 50 of their own, most experts say the French won the

fight for control of the air over Verdun. The supremacy of the Eindeckers was ending, but the fight was bitter. De Rose was killed. Navarre was so severely wounded that he never flew combat again. Georges Guynemer was also wounded. During one of the air fights, Boelcke had the fight of his short life with a Nieuport 17 "Bebe". Outmaneuvered, his Eindecker was riddled with machine gun fire. He barely escaped with his life. In his report, he recommended the Germans develop a biplane fighter and organize in Jastas (squadrons). The German High Command listened and the reorganization mentioned above was ordered. That reorganization would make the Germans truly formidable in 1917 when they again introduced superior fighters. It was interesting to discover that according to Boelcke, the Nieuport 17 was superior to his Eindecker.[57]

Incidentally, the now famous Ace system (five kills today define an Ace) was instituted at Verdun by de Rose to boost morale. He originally designated 10 victories as the guideline which was reduced to five victories when America entered the war. This isolated incident to raise the morale of desperate men flying over Verdun in 1916 became a standard for the world.[58]

Further north a few months later, the Allied supremacy was confirmed over the Somme as the FE-2Bs, DH-2s, "Sopwith One and a

Left: *The SPAD XIII resembled the SPAD VII, but in reality was almost an entirely new design. Powered by a 220 hp geared Hispano Suiza V-8 and equipped with twin Vickers Mg. The XIII was Eddie Rickenbacker's favorite.*

half Strutters", and Nieuports outperformed the Eindecker and the newer Pfalz E.4s and Halberstadt D.2s. The "Scourge" was over and the Germans were swept from the skies during July and August.[59]

The pendulum swung back to the Germans in late 1916/early 1917 as new Albatros and Halberstadt fighters came on line. The Albatros D-III was a better mount than the DH-2s, FE-2Bs, Nieuports, Sopwith 11/2 Strutter, Sopwith Pups and particularly the obsolete BE-2Cs that the Allies were still flying.[60] The German domination peaked in April 1917 in what the Allies called "Bloody April". The fighter forced every other air mission to dance to its tune as the fighter determined who could achieve air superiority. Observation of ground troops, artillery spotting and direction, bombing and strafing in support of ground forces, the interdiction of sup-

plies, and bombing of factories and cities – every air mission, depended on control of the air which was determined by the best fighter. And from January 1917 until May, the Albatros reigned supreme.[61] Coupled with the new squadron (Jasta) and wing (Jagdgeswader) organizations and tactics, and superbly led (e.g.;Baron von Richthofen) the Germans again were dominant.

Then in the summer of 1917, the Allies received a flood of new aircraft (the SE-5, SE-5A, Nieuport 28, the Camel, the F2A/F2B Bristol Fighter, and the SPAD) each equipped with an effective gun synchronizer, and many were armed with two or three guns. Air supremacy returned to the Allies to stay.[62] From 1916 on, the Germans were also outnumbered in the air. This increased by 1918 to overwhelming numbers.[63]

SUMMARY

DOCTRINE, AIRCRAFT PERFORMANCE, STRATEGY, AND TACTICS DOCTRINE

To quickly review the air war: air superiority/supremacy swung back

and forth from the Allies to the Germans based on the superiority of each sides' fighters.

At first, as the fighter weapons system developed, the Allies had maintained a superiority with their Lewis gun-equipped pusher aircraft as they had unencumbered fields of fire versus the lesser equipped German tractor aircraft. Suddenly, the Fokker synchronizer gave the Germans a better weapons system. We then saw the Fokker Eindecker fighter weapons system providing an advantage that led to air superiority for the Germans over the front from mid-1915 until the spring of 1916. Not until improved British pushers (DH-2 and FE-2B) and the French Nieuport 17 appeared in mid-1916 did the tables turn for the Allies. The Allied superiority lasted until early 1917. The Albatros and Halberstadt fighters again reversed the superiority for the Germans peaking in April 1917 in what the Allies called "Bloody April". In mid-1917, the introduction of the Camel, SE-5A, the Nieuport 28, and SPAD, in large numbers, each with synchronized forward firing guns, regained Allied superiority and kept it, until the end of the war.

It clearly was the fighters that determined the tide of the WW I air war by providing control of the sky, and WW I was the precursor of all the wars to come. The air power theorists who were very aware during the "Great War" of the importance of the fighters seem to have forgotten it immediately after the war ended. They became enamored with the strategic bombing theories of destroy-

Below: *The Sopwith Camel shot down more German aircraft than any other Allied aircraft. With its rotory engine it was highly maneuverable. Since it was very light, the torque of the rotory made it a handful to fly at slow speeds.*

ing the war-making capabilities of nations and came to believe that bombers could do the job unescorted and alone. But considering just one WW I example, they should have known better: For six months during the fall and winter of 1915-1916, the Allies could not conduct unescorted observation or bombing missions without unacceptable losses from German Eindeckers. And there were many other examples. How soon dedicated folks forget facts that might upset cherished beliefs.

In mitigation, these theorists were also captivated by the difficulty of detecting and intercepting incoming bombers. The defender had to detect the attack, climb to altitude to intercept, then intercept the bombers in all kinds of weather. This was difficult, if not impossible, with WW I equipment. However, they appear to have been short sighted about the projected development of the fighter. Obviously, no one foresaw electronic detection systems such as radar and that is understandable. But one thing was obvious based on WW I experience: If the fighter intercepted the bomber, the bomber was a dead duck. Further, when you or your opponent had air superiority/supremacy, the odds of successful interception were significantly improved.

It is also very important to note that all during the war that the Allies (largely the British) maintained the strategic offensive, even when they did not have air superiority, and the Germans steadfastly maintained the strategic defensive. The Allies continued to fly into German airspace even when faced with significant German air

superiority, while only occasional German observation planes and bombers penetrated Allied airspace. With the prevailing westerly winds, this gave the Germans a number of advantages as the air battles for them were largely over friendly territory. Allied pilots who were forced down became prisoners while the German pilots were home. With the 100 MPH planes these men flew, Allied airmen found it took a long time to get back to friendly territory against the strong 40-50 mph winds at altitude. Also, as the air battle took place overhead, German ground forces often directly observed them and German confirmation of results was much easier. All of these facts may help account for Allied losses being significantly higher than German.[64]

It is interesting to note that this defensive strategy influenced the German aircraft designers. All during the war, German fighters generally had less fuel capacity than the Allied fighters. Obviously, when fighting defensively over your own territory you need less fuel. Less fuel means less weight and therefore more performance. But range is a key principle of air war. If an aircraft runs out of gas, it is as big a loss as if it were shot down. In the turning fights of WW I, low fuel was often deadly, as the first one to turn for home, exposed his tail. (It has been generally concluded that low fuel allowed Richthofen to shoot down Lanoe Hawker in early 1917.) This predilection of German designers to gain performance by limiting fuel carried over to WW II. During the Battle of Britain, German ME-109Es did not have the range or the time over their targets to properly escort German bombers. They often were forced by low fuel to leave their bombers unescorted, if they were to

make it home to bases in France. Nor was a reliable external drop tank system designed and engineered in time for the key 1940 battles.[65] This error may have cost Germany the victory in WW II.

WW I trained Luftwaffe chiefs were flabbergasted by the long range capabilities of the American fighters of WW II, particularly the P-51. It was incomprehensible to them that a fighter aircraft that could fly to Berlin and back from English bases could outperform their best aircraft. It all started with a defensive strategy originating in the early days of the air war in WW I where low fuel capacity equalled improved performance.

FIGHTER AIRCRAFT PERFORMANCE

Once the air war began in WW I, it was clear that what General Carl "Tooey" Spaatz said years ago was true: "A second-best airplane is like a second-best poker hand, it ain't worth a damn." The rub comes in finding the criterion for the "best" airplane. And since 1914 to date, that debate continues.

After the decision was made to align fighter guns with the axis of the aircraft and to aim the guns by maneuvering the aircraft itself, the flight and maneuvering characteristics of the aircraft became critical to the fighter as a weapon. Obviously, with machine guns fixed to the longitudinal axis of a fighter and designed to fire forward, the vulnerable area was the rear of the aircraft. One fighter maneuvering toward the tail of another fighter was reminiscent of a dog attacking another while avoiding his opponent's fangs which were mounted up front. Fighters, like dogs, must maneuver to attack from the rear and defending fighters like defending dogs must always try to protect their tails. Hence, the derivation of the term for fighter combat in WW I, "Dogfighting". If dogs attacked and defended in three dimensions, the analogy would be exact.

Initially, when WW I air forces examined fighter performance, it was apparent that speed was important as the fighters had to be able to catch the observation and bomber aircraft they were attempting to intercept. As the intruding bombing and observation aircraft were normally at altitude, the fighter's ability to climb to the intruder's altitude was also important. It was also easy to conclude that fighter maneuverability was desirable. Consensus on these points was relatively easy to achieve. But after these points, agreement became more difficult as one aircraft was pitted against another, and it was nearly impossible to build an aircraft that had every desirable characteristic and that could out perform all the other contemporary fighters. It was therefore useful, in many cases, to emphasize or establish design priorities for some of the characteristics, to the expense of others. From that time until now, the argument has continued about what is the most important fighter characteristic: speed or maneuverability.

Given a choice, most pilots seemed to prefer speed, since speed allows the attacker to attack and the defender to successfully run. But as history ably demonstrates, there was no total consensus. Even Baron Manfred von Richthofen chose and flew the slower more-

maneuverable Fokker triplane over faster models. Albert Ball is said to have preferred the nimble Nieuport "Bebe" to the stouter, swifter, SE-5A. Both Aces preferred ships that maneuvered well and were exceptional climbers to the faster, stronger "mounts". Both died in combat while Udet, Fonck, and Rickenbacker, who preferred speedy and more powerful ships survived. Message: energy (speed/altitude) can always be used to maneuver (dive and zoom).

Also a factor in the WW I fighter war was the change from "one-on-one" tactics early in the war to formation fights after 1916. Aircraft attacking singly were vulnerable as they could not attack looking forward and clear their tail simultaneously. Surprise was deadly. Consequently, longevity was not the "loners" strong suit. Most of the survivors benefitted from the switch to squadron and wing tactics of 1917-1918. The renowned loners included Navarre, Garros, Ball, Guynemer, Immelmann, and Frank Luke – all of whom were shot down. The Canadian, Billy Bishop was one of the few "loners" to survive. It follows that he was either very good or very lucky. In the main, these loners seemed to prefer the lighter more maneuverable ships. Guynemer was one exception. He of course, chose the stout, fast Spad. However after close study, there really does not appear to be any clear pattern that emerged during WW I between speed and maneuverability, other than personal preference.

It is also important to recognize that aircraft performance changes with altitude. For example, at higher altitudes engine power may drop off and aerodynamics may change, so perfor-

mance evaluations must take these characteristics into account. For example, at 15,000 feet the Sopwith Pup could outturn the Albatros and maintain its altitude.[66] At lower altitudes, the more powerful Albatros could use its power and energy more efficiently.

This is also true of speed. For example, if the Nieuport 28 could be lured into a high speed dive by an Albatros, it had to pull out very gently to keep from shredding the fabric on the upper wing. With a sturdier fighter on its tail, a gentle pull out would mean almost sure death.

Factors of altitude and airspeed appear again and again in any analysis of fighter performance. For example, the Japanese Zero fighter in WW II was highly maneuverable, but above 230 MPH its ailerons stiffened and became very heavy. The F4U-1D, Corsair, with better high-speed ailerons could roll circles around it at higher speeds. Conclusion: Do not slow down and dogfight in a Corsair versus a Zero. Use your better power and speed to attack when advantageous.[67]

Many other factors were noted as significant in WW I fighter performance by Edward Sims in his books. Most are still significant in today's fighters: (not in order of importance)

Speed

Armament

Sturdiness - how much stress the airframe can stand

Ceiling

Reliability – engine and guns

Acceleration -- key in escaping an attacker and overtaking a fleeing aircraft

Diving velocity – ability to overtake or escape

Rate of Climb

Endurance/Range

Maneuverability

Location – whether over enemy or own lines

Wind – factor in 1914-1918 not much after

Formation – fighting together provides advantages [68]

STRATEGY

When the air strategy of WW I is evaluated, one is almost tempted to say there was none. The air war revolved around air observation in support of ground forces particularly the artillery.[69] It contributed in direct proportion to its success in supporting the army. As the war progressed, the fighter dictated the size of this contribution as it dictated the control of air observation of the various battlefields. In this sense, the strategies that should have been projected by these experiences should have emphasized fighter operations. That did not happen. The horrible stalemates of the trenches and the dominance of the artillery on the battlefield made a strategic bombing solution without such possible stalemates very, very attractive.[70] Although close support was used throughout the war, it disappeared from many nations' national strategies after the war was over. When the Nazis re-introduced close support as part and parcel with "Blitzkriegs" in WW II, the world was in a scramble to catch up. Curiously, the U.S. Air

Service was one of the air forces that maintained a close-support, "Attack Group" between wars.[71]

Of all the Air Forces involved in WW I, the Royal Air Force was the most interesting. Air Marshall Trenchard was one of the great theorists of air power and the only air leader to achieve an individual Air Force organization during the war. (It took U.S. leaders 30 more years to make this progress.) Over half of the German and French Air Services were observation aircraft at the close of the war. Fifty-eight percent of the RAF were fighters. Allied observers found the British pilots "quite sporting" – they treated air war like a game. To the Germans fighter pilots however, the British were seen differently and were described over and over again as "Zahigkeit" (i.e., tenacious). They maintained deep fighter patrols throughout the war often with obsolete aircraft and even when the Germans had air superiority. This often accelerated losses. The French and Germans rarely flew such missions.[72] Finally, they also insisted that their airmen fly obsolete aircraft (the Royal Aircraft Factory insisted the planes were not obsolete) during critical phases of the war (e.g., the BE-2Cs in 1916-1917). In age old English tradition, they lost many battles, and many pilots, but not the war.

TACTICS

The tactics of WW I were basically the same as those in later wars except that they represented the beginning of the beginning. To illustrate this point, WW I started in 1914 and continued through the first part of 1915 as one-on-one tactics. One pilot attempted to attack one enemy machine using the

target aircraft's blind spot if possible. After the Fokker Eindecker appeared (mid-1915), the Germans had a viable weapons system around which to formulate air doctrine and tactics. In Oswald Boelcke and Max Immelmann they had leaders who could outline and teach these tactics. On the Allied side the same process took place in vicious combat. The fighter war came of age over Verdun and was refined over the Somme and the Aisne. The one-on-one combats of the early days were gone in 1917. Like the dinosaurs of antiquity, the "Loners" of the war, Ball, Guynemer, Luke, etc., just disappeared, unable to survive in the new formation wars. By 1917, you fought as a team or you were history. That is still law today, the teams are simply more refined, sophisticated, and bigger.

The same performance rules apply – keep your airspeed (mach) up, never give away altitude (altitude and airspeed equate to energy) and apply your system's strengths against your foe's weaknesses. He who sees or detects the other first, usually wins. Most kills are made when the bandit never sees you or until it is too late. There are new rules to deal with missiles, but even those are much the same. You now depend on electronic sensors as well as your eyeballs.

The fighter and the fighter pilot were keys to control of the air in WW I. We will see in future chapters that many men had to die because lesser men have not learned that lesson. It has applied in every war since and with some slight changes will apply in tomorrow's wars as well – even as fighters become true space vehicles.

NOTES

1. Only a few die-hard fighter pilots felt that the fighter could successfully intercept bomber attacks prior to the Battle of Britain. For example, in the U.S. Army Air Corps, Claire Chennault was retired early as a Captain largely for his adamant fighter views. (See *The Maverick War* by Duane Schultz, St. Martin's Press, New York, p. 1.)

2. The principles of war are a series of corollaries promulgated by military theorists to use as a checklist in planning strategy and tactics. They have over the years become part of military dogma and doctrine. *U.S. Army Field Manuals* list nine principles of war (FM 100-5 dated May 1986, Appendix A):

o Objective — Direct every military operation toward a clearly defined, decisive, and attainable objective.

o Offensive — Seize, retain and exploit the initiative.

o Mass — Concentrate combat power at the decisive place and time.

o Economy of Force — Allocate minimum essential combat power to secondary efforts (usually used to achieve Mass at a more decisive place and time). Parens. added by the author.

o Maneuver — Place the enemy in a position of disadvantage through the flexible application of combat power.

o Unity of Command — For every objective, ensure unity of effort under one responsible commander.

o Security — Never permit the enemy to acquire an unexpected advantage.

o Surprise — Strike the enemy at a time or place, or in a manner, for which he is unprepared.

o Simplicity — Prepare clear, uncomplicated plans and clear, concise orders to ensure thorough understanding.

3. Eugene M. Emme, The Impact of Air Power, Van Nostrand Co.,Princeton, New Jersey 1959, pp. 161-162: Extracts from Command Of The Air by Guilio Douhet.

4. Lee Kennett, *The First Air War 1914-1918*; The Free Press; New York; 1991; pp. 69-70. "More noticeable was the effect of the enhanced fighter force on French bombing missions into Germany. These had been conducted at relatively little cost, but then in a raid against Saarburg on August 2 (1915), the French lost nine planes. It was the beginning of a pattern of losses that would eventually bring an end to daylight bombing." It is clear that, with air power, experience is not always the best teacher. Nor are memories long.

5 .Op. Cit.; p.229.

6 .Emme, Op. Cit.: *"Defeat of the Luftwaffe"*, p. 258.

7 .Op. Cit., p. 227.

8 .Op.Cit., p. 258.

9 .Richard P. Hallion, *Storm Over Iraq, Air Power And The Gulf War*, Smithsonian Institution Press, Washington, 1992, Appendices A-G; pp. 275-317.

10 .Valerie Moolman, *The Epic of Flight: "The Road to Kittyhawk"*, Time Life Books, Alexandria, Virginia, 1980, pp. 19-38.

11 .Ibid.

12 .Ibid.

13. Donald Dale Jackson, *The Epic Of Flight: "The Aeronauts"*, Time-Life Books, Alexandria, Virginia, 1981, pp. 76-100.

14 .Ibid.

15 .Ibid.

16 .Kennett, Op. Cit.; pp. 23-30.

17 .Moolman, Op. Cit.; pp. 41-55.

18 .Op. Cit.; pp.73-100.

19 .Op.Cit.; pp. 107-160.

20 .Curtis Pendergast; *The Epic Of Flight: "The First Aviators"*, Time-Life Books, Alexandria, Virginia, 1981, pp. 146-163.

21 .Kennett, Op. Cit.; p. 18.

22 .Ibid.

23 .Norman Franks, *Aircraft Vs. Aircraft*; MacMillan Publishing Co., New York, 1986, pp. 10.

24 Air superiority is defined as the ability to control a given portion of airspace at least temporarily. Air supremacy is defined as the domination of airspace in an entire theater of operations for longer periods. Once superiority is gained both require continued vigilance and effort to be maintained.

25 .Batchelor and Cooper, *Fighter*, Charles Scribner & Sons, New York, 1973, pp. 1-51.

26 .Ibid.

27 .Kennett, Op. Cit.; p. 104.

28 .Op. Cit.; pp. 104-105.

29 .Jablonski, Op.Cit.; p. 188.

30 .Ibid.

31 .Ibid.

32 .Op. Cit; p.188.

33 .I first met John Boyd, *"the father of Energy Maneuverability"*, at Squadron Officer School in 1957. After a discussion on air tactics, I asked a friend, Capt. Billy Ellis, who was then a USAF Thunderbird, if John was as good as he talked. To

my amazement, Billy said, "Better". About a year later after going through F-100 upgrade training at Nellis AfB, I found out for myself, firsthand. He was the best F-100 driver I ever saw. John went back to school and got an advanced aeronautical engineering degree; then came back, codified and preached energy maneuverability to the pilots in their idiom, and to the engineers in their calculus. Largely through his incredible work after the Vietnam War in the Pentagon, the F-15 and F-16 were built and bought. Unfortunately, John told it like it was and somewhere along the line he annoyed some of the "middle grounders" that inhabited the Pentagon. He always forced people to take a position which bothered some (usually weaker) people. It was travesty that John Boyd was never promoted to general. He was a four-star fighter pilot.

34. This grapnel method of attack was successfully used by a Russian ace Captain Kazakov on March 18, 1915. See *Aircraft Vs. Aircraft* by Norman Franks, MacMillan Publishing Co., New York, 1986, p. 11.

35. Batchelor and Cooper, Op. Cit.; pp.8-9.

36. Jablonski; Op. Cit.; p. 16.

37. Batchelor and Cooper, Op.Cit.; pp.7-8.

38. Ibid.

39. Franks, Op. Cit.; p.11.

40. Batchelor and Cooper, Op. Cit.; pp 10.

41. Batchelor and Cooper, Op.Cit.; p.19.

42. Op. Cit.; p. 10.

43. Batchelor and Cooper, Op.Cit.; p. 10.

44. Edward Jablonski, *The Knighted Skies*, G. P. Putnam, New York, 1964, pp. 34-35.

45. Franks, Op.Cit.; p. 15.

46. Batchelor and Cooper, Op. Cit.; pp. 10-13.

47. Brendan Gallagher, *Ilustrated History Of Aircraft*, Longmeadow Press, Norwalk, CT., 1977, pp. 28-29.

48. Jablonski, Op. Cit.; pp. 36-37.

49. Jablonski, Op. Cit.; pp. 39-45.

50. Jablonski, Op.Cit.; pp.135-136.

51. Franks, Op. Cit.; pp. 26-27.

52. Ibid.

53. Jablonski, Op. Cit.; p.49.

54. Ibid.

55. Batchelor and Cooper, Op. Cit.; pp. 24-25.

56. Ibid.

57. Ibid.

58. Ibid.

59 .Batchelor and Cooper, Op. Cit.; pp. 33-34.

60. Ibid.

61. Gallagher, Op. Cit.; pp.34-35.

62. Konnott, Op. Cit.; pp.72-73

63. Ibid.

64. Edward H. Sims, *Fighter Tactics and Strategy* 1914-1970, Harpers & Row Publishers, New York, 1972, p. 11.

65 .Sims, Op. Cit.; p.11.

66. Batchelor & Cooper, Op. Cit.; p. 26.

67. Alfred Price, *Fighter Aircraft*; Sterling Publishing Co., New York, 1990, p. 96.

68. Sims, Op. Cit.; p. 22.

69. Kennett, OP.Cit.; pp. 220-222.

70. Ibid.

71. Ibid.

72. Ibid.

Above: *Between Wars the equipment steadily improved. The P-6E was one of the most attractive fighters of the early 1930's (1931). Many believe even today that a fighter that looks good will be. The P-6E was built by Curtiss and powered by a 600 hp Curtiss Conqueror engine. Max speed 193 mph, armament was two 0.3 caliber Mg. It still remains one of the prettiest of the biplane fighters.*

CHAPTER 2

FIGHTER DEVELOPMENT BETWEEN WORLD WARS

BARNSTORMERS & SPEED KINGS

When WW I ended, there was a surplus of aviators and airplanes. The WW I aviators were captivated by the sky and were not able to kick the flying habit. Flying, like drugs, creates addicts. These men (women soon joined them) were dedicated to flying and they were determined to make a living doing what they loved. It was largely these "Barnstormers" that kept aviation before the public and inspired new generations to fly.[1]

As the Barnstormers captured the imaginations of the public, it was the "Speed Kings" that not only captured their imaginations but inspired and motivated the engineers to produce aircraft that could fly faster, farther, and climb more quickly to higher altitudes. This transferred directly to fighter military capabilities. As we saw in the first chapter, speed and the ability to climb to altitude both provide energy to outperform and outmaneuver an adversary. Using this "energy maneuverability" a fighter could win a "dogfight" by getting on an opponent's tail. So as the racers raced, the technology transferred directly to build better fighter aircraft. For example, the famous Supermarine Spitfire of RAF WW II fame had ancestry in the Schneider Cup racers.[2]

Racing planes pioneered new technology in streamlining and engine development. In the United States, The Bendix Trophy Race (across the country) and National Air Races (Thompson Cup closed course race) defined many new breakthroughs that later appeared in American fighter design. In fact, in the late 1930s, U. S. Army Pursuit (Fighters) aircraft participated extensively in the races. Earlier in the 1920s, Jimmy Doolittle had flown an Army fighter to win one of the first races. His chief competition was a Navy fighter. Incidentally, the U.S. Navy used the term "Fighter" from 1922 on and it was the Army Air Corps who used the term "Pursuit" (the Navy used the Army pursuit designation from 1919-1922).[3] Other racing planes were built in backyards and garages. But many modern features such as retractable landing gears, constant-speed props, all-metal monocoque stressed skin construction for monoplanes, and landing flaps were developed and perfected in these designs. The racers actually pioneered much of the research and experimentation on increased engine power and cooling, the location of the aircraft center of gravity, and the understanding of high-speed stalls. For example, the Gee Bee airplane was all engine in a low drag, short fuselage design that was inherently unstable, but it was clocked at over 300 MPH in 1931.[4]

The importance of these races to fighter design can be seen in that the Wendell Williams racer was designated the XP-34 and thoroughly evaluated by the USAAC as a fighter in 1935. In the Schneider Cup races, aerodynamically "dirty" float planes (the large floats produce high aerodynamic drag) built by Supermarine (Great Britain) and Macchi (Italy) exceeded land plane speeds by over 100 MPH.[5] The engines that powered these aircraft were specialized to run at max power for only minutes. By the early 1930s, entrants were using large amounts of the aircraft skin surface as flush radiators to cool these unbelievably powerful engines thereby avoiding the drag associated with radiator air intakes. These engines were running on high-energy chemical fuels and producing over 2,600 HP. In contrast, fighter aircraft engines used in the Spanish Civil War between 1936-39 did not top 1,000 HP until the very end of that war.[6] However, the service lives of these "souped-up" racing engines were measured in minutes before they failed and they were not suitable for operational fighter use. The later Italian versions even used counter-rotating props to cancel out the unbelievable torque generated by their designs' powerful engines. The last Italian model flew at 440 MPH. That record still stands for prop-driven float planes.[7]

FIGHTER DEVELOPMENT

By the late 1920s, it was apparent to aircraft designers that fighter biplane designs were reaching their upper limits. This realization took place with bombers and transports first. Part of this was due to Air Marshall Trenchard's influence in Britain, General Giulio Douhet's book, *Command Of The Air*, and Billy Mitchell's successful bombing tests in 1921-22, all of which pushed the bomber to the forefront.[8] However, there were other reasons. Engineering-wise, the design of the stronger, heavier, all-metal monoplanes appeared to be more feasible using the power available from multiple engines. By 1930, many of the world's air forces were working on monoplane bombers, with retractable landing gear. These bombers were faster than the contemporary fighters of their day which were in the main biplanes with fixed landing gear. For example, the U.S. B-9 & B-10 bombers were much faster than contemporary biplanes fighters.[9]

These 1930-1939 bombers/transports were to mislead the strategic air power theorists into thinking they could escape from fighters if they were already at altitude. For example, the Germans thought their Dornier Do-17, Heinkel 111, and Junkers Ju-88s would be able to speed away from many fighters due to their experiences in the Spanish Civil War with slower Republican fighters. The British fell into the same trap with the Bristol Blenheim. In the U.S., the B-17 was conceived to be too fast at high altitude to be successfully intercepted. These Air Power theorists were comparing their bombers with contemporary in-service fighters. But fighters were then behind. When Hitler Germany began its saber-rattling and WW II loomed on the horizon in 1935-39, fighter design exploded and the bomber was caught and surpassed in performance . By 1939, new fighters in Great Britain and Germany had exceeded bomber performance in terms of speed, by 50-100 MPH and this was just the beginning.[10]

The transition years from biplane to monoplane fighters were educational as they appeared to show much about human beings' resistance to change. Many countries began to change to monoplanes in the early 1930s, but some countries (largely based on pilot preferences) continued to build biplane fighters. Italy is a prime example and Russia also continued to refine and improve its I-15 right into WW II. England, (Gloster Gladiator), Italy (CR-42), and Russia (I-153) all flew biplane fighters in WW II.

The U.S. Navy was also slow to convert to monoplanes because of strength concerns related to carrier fighters. It was believed that landing and arresting stress loads were too high for monoplanes. The primary navy fighter in 1939 was the retractable gear Grumman F3F which many consider the ultimate biplane fighter. All biplanes fighters were gone from U.S. carriers by 1940.

In Japan, interest in carrier qualified monoplane fighters began in the early 1930s and the Type 96, Mitsubishi monoplane was approved for production in 1936 when the U.S. Navy was still buried in biplanes. The Mitsubishi A5M4 with fixed landing gear came close to the performance of contemporary foreign fighters with retractable gears in the mid-1930s. Japanese fighters were characterized by good speed, excellent maneuverability, long range, and outstanding rates of climb. This emphasis on light construction, modern design, and high maneuverability carried on into WW II.[11]

In 1937, the specification for a long-range shipboard fighter with retractable landing gear was submitted by the Japanese Imperial Navy. In 1940, the Type 0 entered the Fleet. The famed "Zero" was on-board. It was faster, longer-ranged, and more maneuverable than any shipboard fighter in the world and remained so until late 1942.[12]

But in mitigation, it must be remembered that the Japanese were at war in China during the mid-1930s. This war provided an impetus to improve Japanese aircraft not present in other countries. Incidentally, the Japanese fighter aircraft fought with the best foreign aircraft of their era, primarily American and Russian all used by the Chinese Air Force. The Type 96, Mitsubishi; Type 97, Nakajima; and Type 0, Mitsubishi all outperformed these aircraft (Russian I-15s, I-153s, and I-16s; American Curtiss 75s and the export versions of the Seversky P-35) until Chennault's Flying Tigers showed up in 1941. The Americans, Dutch and British had problems with these same Japanese aircraft until 1943. Chennault, and the Russians, learned not to dogfight with these Japanese fighters. The rest of the world would learn this lesson in WW II.

In fact, however, much of the world's pre-WW II fighter development must be attributed to Germany.[13] Not only did the Me-109 dazzle the world, but when Hitler started his air build-up in 1933, all of the nations of Europe followed. The successive invasions of the Rhineland, Austria and Czechoslovakia spurred British and French development and even awakened the sleeping giant, the U.S.[14]

The Germans, Russians, and Italians took advantage of the Spanish Civil War 1936-1939 to test their best equipment. Initially, the Republicans of the existing government flying Russian fighters had an edge in equipment. The biplane, gull-winged Polikarpov I-15 "Chaika"(Gull) matched up well with the German Heinkel He 51, and the Italian Fiat CR-32, both also biplanes. The Russians then introduced the first monoplane fighter of the war, the Polikarpov I-16 "Moska" (Republican) or "Rata" (Nationalist) which dominated the skies until the advanced German monoplanes, the Me-109, and HE-112 were tested later in the war.

During this war, the need for increased fighter armament became evident. This trend would continue into and during WW II. Remember it is not the aircraft, but the aircraft weapons system that is important in war.

On this side of the Atlantic, during the 1930s, the U.S. lagged behind in fighter development and was far behind the Germans and British. The best U.S. fighters, the P-36 and the P-35, in operational configuration, could not exceed 300 MPH while the Me-109, Hurricane, and Spitfire were much faster, more heavily armed, and just as maneuverable.[15] For example, on June 3, 1936, the RAF ordered 600 Hurricanes, (325 MPH top speed) and 310 Spitfires (360 MPH top) both armed with eight .303 machine guns mounted in the wings. Two weeks later, the U.S. Army Air Corps ordered 77 Seversky P-35s armed with two guns synchronized to fire through the prop, and with a 281 MPH top speed.[16] It appears the USAAC was still fighting WW I.

It was Hitler's aggression that finally impelled the U.S. to catch up with the rest of the world in fighter development. In November 1938 worried about possible war in Europe, President Roosevelt proposed a mega-expansion of U.S. air power, that assured modern U.S. fighters by 1941-42.[17]

AIRCRAFT DESIGN

A quick study of WW I and post WW I aircraft designs enunciates the early designers' problems. These designers were aware of the lower airframe drag of monoplane designs (e.g. Bristol, Fokker, and Morane-Saulnier monoplanes dur-

---

Below: *The Seversky P-35 was ordered in 1936, two weeks after the Spitfire and Hurricane were ordered. The P-35 had two guns synchronized to fire through the prop. Both British fighters had eight wing guns. The P-35 could not exceed 300 mph in level flight. Foreign fighters were superior to the P-35*

ing the war). But fixed engines of higher power were heavy, and with the materials then available the biplane had advantages in strength, durability, and reliability.

Monoplane designs had a checkered career during WW I. The Fokker Eindecker was one of the most successful airplanes of the war. Why this successful monoplane design was not redesigned and improved after 1916 by the Germans remains a mystery. But many monoplane designs of this period had problems with strength. Also the cantilevered wings and control surfaces had to be externally braced with wires and braces that increased drag almost to the level

of the biplane. There were other problems. The Morane Saulnier N was fast and maneuverable but was unstable and landed fast which made it unpopular with pilots.[19] One of the reasons WW I monoplanes were not more popular was pilot prejudice. Even fast and maneuverable fighters get bad reputations if they come apart in the air or if they are difficult to handle and land. Once gotten, bad reputations die hard.

In 1918, the German Junkers Company developed the first all-metal, monoplane fighters and fighter bombers. Designed by Hugo Junkers, the JU-D.I and JU-CL.I had internally braced wings that were covered with an all-metal corrugated skin which took part of the stress from the internal airframe. No external wires or bracing were required. Like an accordion, the corrugated metal structure was very strong

along the corrugated metal ridges, but had to be internally braced to handle other loads.[20] These clean aircraft were very fast, maneuverable, strong, and generally first-rate. Metalworking and production problems kept them from being more successful. The German aircraft industry was not used to metal work. Only about 90 total aircraft were produced before the end of the war.[21] Junkers used steel on these aircraft and weight was a problem. These ships were just too heavy. In 1919, Junkers in his design of his F-13 airliner, substituted duralumin alloy for steel, which was as light as aluminum, but twice as strong.[22] Why others did not build on these breakthrough monoplane designs for future fighters is not known.

Another German designer, Adolph

Rohrbach improved on the Junkers ideas when he hit on stretching duralumin sheets over a light but strongly rigid framework. This structure could handle loads in every direction and was smooth and streamlined on its outer structure. The boxlike inner structure and tightly stretched skin required no internal bracing as the stressed skin of the wings and fuselage shared the aircraft loads, saving weight and space.[23]

In the mid-1920s, when sufficient engine power was available, designers began to use metal to provide the strength for monoplane bomber and transport designs. Not surprisingly, the Junkers company was in the forefront with a number of corrugated metal designs including the rugged all-metal, tri-motor, the Junkers JU-52. The JU-52 served as a transport and bomber during the Spanish Civil War and with the Luftwaffe all through WW II.[24] The Ford Tri-Motor, designed in America, was of the same construction and equally successful.[25] Americans had digested the European progress and were ready to add some wrinkles of their own. Progress was swift.

Anthony Fokker, now a naturalized American citizen, tried to do with wood and fabric what Junkers and Rohrbach were doing with metal. Fokker combined wood and fabric wings with welded steel tube fuselages and produced several sound monoplane transports.[26]
A young American designer, named Jack Northrop, working for a new American company named Lockheed, combined Fokker's wing design with Rohrbach's fuselage and came up with the first truly

streamlined cantilever monoplane. The Lockheed Vega could carry four passengers and cruise at 135 MPH. When combined with a NACA cowl over its radial engine, this cruise speed went up to 155 MPH.[27]

Northrop left Lockheed in 1928 and immediately designed a beautiful low-wing monoplane model named Alpha. Alpha used Rohrbach's stressed skin throughout.[28] It was a modern design and an ancestor of a number of Northrop and Douglas attack and fighter aircraft to come. Alpha was sold to Trans World Airlines. In Seattle, the Boeing company was also using Rohrbach's ideas and added another new wrinkle of their own: a retractable landing gear. When speeds got up to 130-200 MPH, and with drag increasing as the square of the speed, retractable landing gear were now very worthwhile.[29] The Boeing Monomail was followed very rapidly by the Boeing 247 and the Douglas DC-2, DC-3. All were stressed skin, two engined monoplanes with retractable landing gear, NACA nacelles and cowls, and cruise speeds better than 150 MPH. Modern airliners were here. On March 20,

1932, Boeing first flew a stressed-skin low wing fighter, model 248, which became the famous P-26, "Peashooter". A total of 136 were accepted by the U.S. Army Air Corps in the summer of 1934. The P-26 still had fixed landing gear. Fifty P-30 two seat fighters were ordered in December. The Consolidated P-30 was the first USAAC fighter to have a retractable gear, turbo-super-charger, and be a monoplane design. Two years later when 77 P-35s were ordered, the next generation of American fighters was ushered in.[30] The fighters had finally caught up with bomber and transport design.

In England, 600 eight-gun Hurricanes and 300 Spitfires were ordered into full production two weeks before the USAAC ordered

---

Below: *One of the fastest 1930 racers was the Gee-Bee. The aircraft was short, stubby and practically all engine. The short stubby fuselage and large engine induced serious directional stability problems, but it was fast.*

the P-35. The German Me-109 (the correct title was Bf-109, but I will use the more popular Me-109) began its life a few months earlier. All of these European fighters used liquid-cooled, in-line engines and the slim, pointed-nose, European fighters were significantly faster than contemporary radial engine powered American fighters. These fighters powered by liquid-cooled engines were faster because they had less frontal area than the blunter-nosed radial engine fighters and therefore had less parasitic drag even though the in-line engines were heavier. American designers quickly produced fighter designs based on the new Allison V-12 in-line engine built by General Motors. The results were the P-38, P-39, P-40, and later the P-51, all at least partially successful in WW II.[31] The P-35 design was to grow into the P-43, and then the P-47 which used bigger, more efficient radial engines in new, lower drag cowls.

Between 1936 and 1939, the low-wing monoplane came to dominate fighter design and these new swift fighters gained a clear ascendancy over the biplanes of the day. The I-16 Moska and Messerschmitt Me-109 dominated the Spanish Civil War.[32]

The Japanese Type 96, 97, and 0 dominated the Sino-Japanese War

until the P-40s of the American Volunteer Group appeared.[33] The Type 96 and "Zero" were designed to take-off and land on aircraft carriers. Aircraft that operate from aircraft carriers must land and takeoff at very slow speeds and be structurally very strong to withstand the high impacts of short-field landing and cable arrestment. Biplanes were admirably equipped to meet these requirements. However, aircraft performance requirements dictated that navies of the world develop "carrier qualified" monoplanes. With the Type 96, the Japanese were first in the world to develop monoplane carrier fighters. Uncharacteristically, the Type 96 outperformed many land-based fighters of its day, even with a fixed landing gear (many contemporary land-based fighters had retractable gears). Usually the higher structural requirements of the carrier fighter (higher

strength usually meant more weight and less performance) mitigated the carrier fighter's performance. This was not true of early WW II Japanese carrier fighters. Neither the Japanese nor the U.S. Navy would enter WW II with any operational biplane carrier fighters.

The U.S. Navy ordered its first monoplane fighter the F2A-3, Brewster Buffalo on June 11, 1938 and deliveries began June 1939. However, when this aircraft was modified to carry armor and self-sealing fuel tanks, performance fell dramatically. With a small wing, it appeared to be impossible to modify the Buffalo design to meet WW II weapon systems demands. In its first WW II battles in Burma, the Dutch East Indies, and at Midway, the Buffalo was shot from the skies

by the superior Mitsubishi Zero.[34]

It remained for the aircraft that lost to the Buffalo in the Navy fly-off in 1938 to save the day in WW II. The Grumman F4F, Wildcat became the Navy's standard fighter from 1940 until 1943. Flying the Wildcat, Navy and Marine fliers were successful in fighting the famed Zero by capitalizing on the F4F's diving ability, firepower and ruggedness.[35]

AIRCRAFT ENGINE DEVELOPMENT

Engine development was one of continuous refinement between wars with no revolutionary developments until WW II. The heavier in-line stationary engines always preferred by the Germans in WW I continued to grow in favor with designers as they

provided slimmer more streamlined fuselages. It should be noted that the Germans preferred the six cylinder in-line blocks while the French, Americans, and British generally chose the V-blocks of eight and 12 cylinders.[36] The engines steadily improved. The American Liberty V-12 engine of 1918 weighed 882 pounds and produced 420 HP. The Rolls-Royce Merlin in the 1938 Spitfire weighed 1,335 pounds but produced a 1,030 HP. Power to weight ratios just about doubled between wars.[37]

Rotary engines, which were limited in power by their immense torque, slowly disappeared from use. They were replaced by fixed radial engines where only the crankshaft and propeller shaft rotated. Retaining the air-cooled features of the rotary these engines remained significantly lighter than liquid-cooled engines. They were also much less susceptible to battle

damage as there was no coolant to leak. However,they were bulkier in circumference and caused more drag than the slimmer in-line and V-block engines.[38]

But thanks to the work of Frank Weick at the U.S. National Advisory Committee for Aeronautics (NACA), this drag was reduced substantially. Prior to Weick's work, designers had left radial engine cylinders exposed to the air. Since the engine was air-cooled this seemed to make good sense; however, Weick discovered that drag could be substantially reduced by enclosing the engine in a smooth cowl. There was no reduction in cooling if the cowl was open to the air in front and back. He also found that multi-engine aircraft gained similar benefits if their engines were cowled and placed in nacelles that were faired into the wings. Cowled engines suspended below the wings as in the contemporary Ford Tri-Motor gained no benefits.[38]

Left: *In the 1920s, the Curtiss "Jenny" was used by many barnstormers and many fighter pilots first flew in this venerable old biplane.*

The NACA cowls and nacelles were very important to engine development in the U.S. as American engine manufacturers Pratt-Whitney and Wright had assumed a prominent role in the development of radial engines worldwide. And with this discovery, the lighter, potentially more powerful radials could successfully compete with their liquid-cooled competitors.

Aircraft engines continued to improve in reliability between wars and steadily improved in power. Some of this increase in power came from improvements in fuel. Octane ratings (resistance to premature detonation) steadily improved due to the American discovery of tetra-ethyl lead. By the 1940s, U.S. aircraft were routinely using 100 Octane fuels. These better fuels allowed higher compression ratios, better fuel economy, and more powerful engines.[39] It is important to note that the Germans had no access to these improved American fuels. There were also significant improvements in metallurgy which improved power-to-weight ratios even more.

One of the most significant developments was the advent of supercharging. As altitude increases, the density of the air decreases. At the surface (mean sea level) the density of the air that surrounds us is about 15 pounds per square inch. At 7,000 feet it is half the density of sea level. As reciprocating aircraft engines burn fuel at a constant ratio with the oxygen in the air, an unsupercharged engine can produce only about half as much power at 7,000 feet as it can at sea level. Engine power depends on how much fuel we burn and the amount of air we

can pump through the cylinders determines how much fuel we can effectively burn. By adding an air pump (supercharger) to the intake of the engine we can produce sea level or higher air densities inside the engine even at altitude.[40] When you consider air densities of only a few pounds per square inch at 35,000 feet, you can see the importance of superchargers to engines at high altitudes.

There were two types of superchargers for these engines: internal and turbo. The internal superchargers were powered directly from the engine driveshaft (using some of the available power) usually in one or two stages to keep from over-boosting the engine. Turbo-supercharging on the other hand, uses a turbine in the exhaust flow of the engine to compress the air. Since the exhaust gases are normally wasted, the turbine energy to compress the intake air is practically free. The problems with turbo-charging, however hinged largely on metallurgy. It proved very difficult to develop turbines that could sustain very high revolutions per minute (RPM) in the hot exhaust gases of powerful aircraft engines. The American General Electric company pioneered the development of heat-resistant turbines for turbo-chargers. This research and development later helped GE develop some fine jet engines.[41]

The emphasis on supercharging can be more easily understood when it is realized that because of the decrease in air density, airframe drag is proportionately reduced at altitude, therefore, with the same power, true airspeed (indicated airspeed corrected for density) and

ground speed (speed in relation to the ground) increase substantially at altitude. Also in air fighting, the power available from the engine determines much of the energy available to maneuver. If two opposing fighters met at 25,000 feet, the one with the best supercharger usually won. During WW II, the P-51, Mustang was an also-ran fighter until the Rolls Royce Merlin 61 engine was installed. The Merlin had a powerful, two-stage internal supercharger while the original Allison had none. With the Merlin, the P-51 was superior to opposing Me-109s and FW-190s at altitudes from 15,000-30,000 feet where the fighting usually took place.[42]

When designers became convinced of the efficacy of liquid cooled in-line engines in the late 1930s, the only available in-line American engine was the Allison. Because of production difficulties with turbo-superchargers, American designers at Bell (P-39) and Curtiss (P-40) had difficulty in providing successful supercharging for this engine. These models consequently had problems performing at higher altitudes against Axis models. Below 12,000 feet they were competitive, but at higher altitudes the engine could not perform.[43] The Lockheed P-38, with two Allison engines was delayed for two years when the only prototype was destroyed on a cross-country stunt. By the time this aircraft became operational it was hoped that turbocharger teething problems had been largely solved. This seemed to be proven true when the aircraft were assigned to North Africa during WW II, and there were very few engine prob-

Top: *The Grumman F3F was the U.S. Navy's top fighter in 1939. It still remains one of the top biplane fighters of all time.*

lems. But when the P-38s started flying at altitude over Northern Europe, they began to have engine problems. For example, one problem was with the turbocharger regulators which would freeze and pilots could pull only 10 inches of intake manifold pressure or 80 inches of pressure. The former provided engine idle power, while the latter would blow the engine.[44]

Remember, other than engine power, the easiest way you could gain additional energy in a fighter is to convert altitude (potential energy) to speed (kinetic energy) by descending or diving. Therefore when at the same speed, a fighter at a higher altitude has an energy advantage over his adversary. When at the same altitude and speed, the aircraft with the most available engine power and/or the ability to climb should prevail. Engine power at altitude is the key to both and between world wars supercharging became the key to engine power as the alti-

tudes of engagement escalated into the stratosphere.

Fortunately, the U.S. as a leader in commercial aviation had concentrated on the building of powerful, easily maintained, radial, air-cooled engines. When WW II broke out in 1939, both Wright and Pratt-Whitney had operational radial engines with multiple radial cylinder rows and the most powerful of these engines already produced around 2,000 HP.[45] Liquid-cooled engines of this era topped out at around 1,300 HP maximum. By the end of WW II, radial engines were up to 28 cylinders and around 3,500 HP while the liquid-cooled engines were around 2,500 HP maximum.

The radial, air-cooled designs gave a clear power-to-weight advantage to its users. As the radials rapidly increased in power, this advantage offset the higher drag caused by the bulkier noses that designers were so concerned with in the 1930s. In addition, the radial had no large liquid-

filled radiator to increase its vulnerability. A hit in the radiator of an in-line fighter was a killer. Once the coolant was gone so was the aircraft. Conversely, radial-powered fighters were legend in their ability to take punishment and still come home.[46]

With engine improvement, also came improvements in propellers. All of the power of these engines had to be transmitted to the air by the propeller and improvement was needed. To absorb increasing engine power, props increased in size, switched from wood to metal, then from two-blades to multi-blades (three or four bladed). The radii/diameter of propellers was limited along with revolution speed as the outer tips of props lost aero-

dynamic efficiency as the tips approached the speed of sound (supersonic aerodynamics were not then understood). But improvements raised the efficiency of prop aerodynamics to around 88%.[47]

One of the problems with props, was matching the reciprocating engine torque curves with prop revolutions. For example, on takeoff, the engine needed to produce maximum power by producing maximum revolutions (Torque x RPM = Power). When cruising where economy was needed, the prop needed to produce maximum thrust for the least number of revolutions. Since the propeller is nothing but an airfoil that rotates, the prop pitch (angle of attack of the prop) determined the load on the prop and to some extent the rotational speed capability of the engine-

prop combination. Therefore, for takeoffs, the prop needed to be at a very low (fine) pitch to facilitate maximum RPM and to produce maximum engine power. For cruise, prop pitch needed to be higher to produce optimum thrust for minimum RPM (minimum RPM meant minimum fuel consumption). Early fixed propellers were a compromise between these requirements.[48]

Between wars, designers realized that aircraft performance could be improved by variable-pitch propellers. The airline builders in particular were interested in improved cruise performance, but the aircraft still had to takeoff. With the prominence of airlines in the U.S., it is not surprising that some of the earliest and best variable pitch systems were developed in the U.S. and copied or licensed by other countries.[49] Hamilton Standard (hydraulic) and Curtiss (electric) systems were among the best in the world. These systems

allowed the pilot to adjust his prop from the cockpit to hold a constant speed given enough engine power.[50]

Interestingly, the Seversky P-35 was inferior to its contemporary British Spitfire when they were ordered, but the P-35 was equipped with a variable-pitch, constant-speed propeller and the Spitfire had a fixed prop. In fact, early in WW II, the Me-109 had much the same advantage over the Spitfire as the P-35. The Me-109 had a manual, variable speed prop. The American constant-speed prop however, was superior to the Me-109 as it was fully automatic. The German prop needed to be manually reset when changing pitch. With fixed props, the Spitfires were really behind. Installation of a two-speed prop raised the Spitfire's service ceiling by 7,000 feet, but even this was not done until 1940.[51]

GUN IMPROVEMENTS

Following WW I, gun designers in Europe, Japan, and the U.S. sought to improve aircraft armament. Efforts were made to develop higher caliber weapons, with higher rates of fire, lower weights, reductions in size, and lower recoil.[52] Higher caliber weapons were sought to reduce the number of hits required to shoot down a target aircraft. The goal was to score one hit and get one kill.[53] During WW II, Spitfires and Hurricanes sometimes scored thousands of .303 caliber hits on German bombers before getting a kill. As aircraft gunnery systems provided only momentary tracking solutions, higher rates of fire were required to maximize probability of hits. Guns are no different than any other airplane component in that weight limits aircraft performance. The lighter the guns are, the better. Aircraft structure must absorb gun recoil; therefore any reduction in recoil will reduce aircraft structural requirements which is, of course, desirable. Finally, designers tried to reduce the size of the weapons to reduce weight and to fit them into the cramped spaces available in aircraft. This became really pertinent when trying to squeeze a number of guns into the thin wings of these fighters.

AIRCRAFT ARMAMENT INSTALLATION

At the end of WW I, the standard armament of fighter aircraft was two machine guns firing through the propeller. During most of the years between the world wars there was very little change. One change that did take place however, was the location of the guns. Advancements in electrical technology allowed guns to be remotely located in the wings and electrically fired from the cockpit. These guns required no prop sychronization as they were installed to fire outside the propeller arc and were set to converge at a given range. Designers were required to design the wing structure to provide for guns, ammunition, and feeder mechanisms. One problem with wing guns was that the breech mechanism and ammo had to be located at or near the center of gravity to maintain control as ammo was fired. The ammunition was heavy and with high rates of fire there was a rapid rate reduction in aircraft weight.[54] Bad weather and the cold weather associated with high altitudes gave these wing gun installations a lot of problems, usually with feed mechanisms.[55]

As late as 1939, many countries still had the two rifle caliber guns in the nose as their standard. In Japan, her frontline Army fighter, the Ki-27, was equipped with only two 7.62 MM guns as late as 1939.[56] Italian fighters had only two, and American fighters ordered in 1936 had only two[57] synchronized guns to fire through the prop.[58]

In the U.S. Italy, and Japan, there were experiments with the replacement of 7.62/7.9 MM (.30 caliber) guns with 12.7 MM (.50 caliber) ones. Both Italy and the U.S. adopted 12.7 MM nose guns as standard. When Russian I-15 and I-16 fighters showed up in the Spanish Civil War (1936-1939) with four 7.62 MM guns,[59] the world took note and began to add guns to the wings.

To add and emplace guns in the wings, after the wings had been designed, was a challenging proposition. For example, retractable landing gear usually folded into the wings. In addition, in the mid-to-late 1930s, the best power plants were large, liquid-cooled engines. To keep frontal areas small and reduce airframe drag, radiator and oil cooler intakes were placed under the wings and cooling air ducted back to the engines through the wings/wingroots (Spitfire, Hurricane, Me-109). In many cases, to retrofit guns into the wings required a complete redesign of the wing.[60] From 1936 on, the trend was to increase firepower as much as possible.[61]

Another problem with machine guns in the wings was that they were mounted to converge at a given range and were only concentrated at that range. Bullet patterns were necessarily dispersed at all other ranges. Of course not having to be synchronized, wing guns could fire at their maximum rate of fire.

But it was the British, without combat experience, that launched the real escalation in fighter firepower when the specifications from the Air Ministry in the mid-1930s demanded eight-gun fighters. The result was the Hawker Hurricane and the Supermarine Spitfire designs ordered in quantity in 1936. Eight .303 caliber machine guns were mounted in the wings of each. These were the fastest best armed fighters of their day.[62]

The Spitfire wing design was very interesting. The symmetry of the beautiful elliptical wing was also functional. The curved wing, though difficult to manufacture, aligned the eight breeches on the guns while providing space for the landing gear, provided maximum wing area within a given wingspan, and provided the maximum chord for attachment to the fuselage. The

alignment of the breeches also simplified weapons feeds, and wing supports for the guns. Incidentally the wing was extremely thin (thinner than the Me-109, and Hurricane). This thin wing showed Reginald Mitchell's clairvoyance as a designer for as aircraft performance neared the speed of sound, thin wings became extremely important.[63] Follow-on Hawker designs (e.g., the Typhoon) with thick airfoils were to

encounter severe compressibility problems at altitudes and speeds that did not bother the Spitfire.

The venerable Me-109 was designed with two nose guns firing through the prop. After experience in Spain and the news of eight gun British fighters, a 20 MM cannon designed to fire through the propeller hub was included. When the Germans had trouble

Left: *Spitfires were improved throughout WWII. Engine power was increased and wing platforms changed based on specific missions starting with two-bladed fixed props on the MkI and ending with contra-rotating props on the MkXXI. Note the four bladed prop on this MkIX.*

with this hub cannon, they installed two lightened 20 MM cannon in the wings in the Me-109E. The Me-109E was the Messerschmidt version that fought the Battle of Britain in 1940. In the small, lightweight Me-109 finding room for wing guns was a constant problem (the feed mechanism on the "109E "Emil" ran from gun to wingtip to wingroot and back). No Me-109 model after the "Emil" ever again used wing guns.[64] The original design had called for nose guns and wing gun designs for the complex, small 109 wing never proved satisfactory.[65]

In February of 1939, a secret American fighter flew across the U.S. in record time only to crash upon attempting to land in New York. The two engined XP-38 was designed with an armament cluster in its nose which had a 37 MM cannon, two 7.62 MM (.30 caliber), and two 12.7 MM (.50 caliber) machine guns. Concentrated, these guns were outside the propeller arcs of this powerful fighter and with no need to converge like wing guns, provided a concentrated column of maximum firepower out to a range

of 1000 meters. The armament package was later standardized to a 20 MM cannon and four .50 caliber Browning machine guns. This got rid of the heavy, slower-firing 37 MM and provided significantly more ammo for the cannon.[66] This armament package was similar in concentration of fire to later jet aircraft. In fact, old P-38 pilots felt right at home in the first operational USAF jet as the noses were nearly identical.[67]

In 1937, Bell proposed a low-wing monoplane with unusual armament. The Bell P-39 had an Allison engine mounted over the center of gravity just above the wing, and driven through an extended drive- shaft. This made room for a heavy 37 MM gun to fire through the prop hub. This cannon was matched with two .50 caliber guns in the nose and four .30 caliber guns in the wings. The airplane was poor at altitude because of no supercharging, but the Russians loved it for low-altitude, close support work. Its heavy armament proved devastating against ground targets.[68]

The Curtiss P-36 grew into the P-40 which finally standardized on four

or six .50 caliber machine guns in the wings. This was usually the armament for the P-51 and all the Navy fighters during WW II. The Seversky P-35 grew into the Republic P-47 which mounted an awesome eight .50 caliber machine guns in the wings from day one.[69]

AIRCRAFT TACTICS

After WW I, much of the fighter tactical knowledge gained during the war was either forgotten or discarded. Line abreast formations were forgotten and concentration was put on bomber interception. Cross-over turns that the RAF had taught in 1922 disappeared.[70] The simple dive and pull-off attacks of the war were forgotten. The RAF settled on the three ship "Vic" as its standard fighter formation. This formation was sound for weather penetration and provided a good lookout in clear weather but proved inflexible in maneuvering fighter combat.[71] Incidentally, the RAF still used the Vic at the beginning of WW II and as late as the Battle of Britain.

The RAF 1922 Training Manual emphasized three fundamental principles: 1) Offensive spirit is necessary for victory; 2) Every attack must be pressed home with the implacable determination to destroy the enemy; 3) Surprise must be employed whenever possible.[72] This does not reflect the depth of Boelcke's Dicta, but the emphasis on surprise is on the mark. The largest percentage of kills in every war has come when the defender does not see the attacker until it is too late.

Between wars, fighter units practiced close formation and aerobatics. It was generally thought that dogfighting was dead due to the high "G" forces caused by maneuvers at the higher speeds. The three ship formation was the standard in many European air forces including the Germans. In light of WW I experience, this appears astonishing.[73]
The biggest improvement in coordinating fighter units was the installation of radio in the aircraft. This single innovation made real cooperation possible on a scale unknown before radio.[74]

German fighter tactics improved dramatically in Spain during the Civil War.[75] Unfortunately for the Spanish Republicans and later for the Allies, the Luftwaffe came up with another Boelcke in Werner Molders who led the Legion Condor in victories (14)[76] and in developing tactical fighter doctrine and tactics. Originally using Vics like the rest of Europe, the Germans developed (or rediscovered) two and four ship formations. Faced with a shortage of Me-109s (they often only had two operational) when the aircraft was field tested in Spain, the Germans adopted a two-ship line abreast formation with about 200 yards separation as their basic element (Rotte). They then included a second Rotte to form a "Schwarm" or four-ship line abreast formation. It was immediately apparent that this formation was far superior to the Vic.

The only remaining problem with this formation was how to maintain position in turns without losing the aircraft on the outside of the turns as the aircraft were flying at maximum or near-maximum power and the outside aircraft had much farther to fly. To cure this problem, Molders instituted the cross-over turn from WW I (and the 1922 RAF training manual!). By cutting across to the inside of the turn the outside ships could fly the same distance as the others. The aircraft simply crossed behind the leader and as those on the outside cut across to the inside, those on the inside slid across to the outside. By playing the turn and taking vertical separation, (to prevent collision) every aircraft could fly the same arc as the leader, making major power changes unnecessary. With practice, everyone in the Schwarm was mutually supporting, even in turns. The leader could also keep his airspeed (energy) up.[77]

Using this formation and technique also provided outstanding lookout on each individual aircraft's tail. It provided all-around look-out capability, and allowed the wingmen to maintain position even at near-maximum power settings and speeds, which is crucial if the formation is to maintain an energy capability to perform maximum maneuvers. The leaders could concentrate on attacking while the wingmen protected their leaders' tails. This formation revolutionized fighter warfare and with very minor variations is still used today.[78]

The Russian pilots opposing the Condor Legion Germans quickly saw the facility of the Rotte and Schwarm and copied the German tactics. They did adjust the tactics slightly by closing up a bit since they had no radios.

When they returned to Russia they recommended the adoption of the four-man flights. They were not adopted due to stodgy, bureaucratic communist leadership and the Soviets had to re-learn the hard way during WW II.[79]

The British and French continued with the three-ship Vics and sustained heavy losses in the first two years of the war.[80] The Italians, even though they were associated with the Germans in Spain, were happy with their maneuverable biplanes and continued individual combat.[81]

Like the Italians, the Japanese promoted fighter maneuverability and they also liked individual combat. It seemed to be in their ancient Samurai tradition which promotes individual inner strength and one on one combats. Unlike the Italians however, they rapidly changed over to modern low-wing monoplanes. But they insisted the monoplane be as maneuverable as the biplane thereby gaining the speed of the monoplane and retaining the dexterity of the biplane. This was done by providing extremely light designs that were modern in every respect. When the Nakajima Type 97 and Mitsubishi Type 96 were flown in individual combat against the best contemporary fighter of its day, the Russian I-16, they held their own. If the Chinese or Russian pilot was foolish enough to dogfight with these nimble, swift, more maneuverable aircraft, they were dead. It can be argued however, that these light maneuverable fighters were not the best type of fighter weapons system. By concentrating on maneuverability, the Japanese forfeited speed and armament. The former, provided the initiative, the latter, the killing power.[82]

Claire Chennault was an interested observer of these battles over North China and Manchuria. As a retired U.S. Army Air Corps fighter pilot, Chennault was hired by Generalissimo Chiang Kai-Shek at $15,000 a year as a fighter consultant and given the rank of Colonel in the Chinese Air Force. As earlier indicated, Chennault was a maverick and an all-out fighter adherent in a bomber dominated U.S. Army Air Corps (USAAC) in the 1930s before he retired. Truculent, opinionated, and intolerant, Chennault gave little truck to those who did not agree with him and very few in the USAAC did, including Hap Arnold, the Chief of Staff.[83] He was not liked by most; however, when history tested the ideas of this great fighter tactician, he was proven right, not his critics. He finally got his chance in 1937 when he retired and was hired by the Chinese.

Chennault had studied the Japanese for years. When he and the Chinese talked the U.S. into providing 100 Curtiss P-40Bs and over 100 pilot volunteers, Chennault was ready. The American

Volunteer Group (AVG), "The Flying Tigers" were born. Weeks before Pearl Harbor, Chennault began to train the AVG in Burma where he passed on the priceless knowledge he had gained. His tactics bore a marked resemblance to the German Rotte and Schwarm, but they had been developed independently. In intensive training in Burma, he briefed his pilots on the strength and weaknesses of the Japanese Air Force – their tactics, their penchant for dogfighting, and maneuverability, the strength of the P-40 versus their fighters, including the "Zero", which at that time was largely unknown.[84]

He pointed out the success of the Russian I-16s using dive and zoom and the effects of their heavier armament. He showed the pilots how the P-40 could employ these tactics and directed their practice in the air. He taught and emphasized air-to-air gunnery and drilled into his pilots that a fighter is a weapons system and only as good as its guns. He forbade dogfighting and turning engagements with the Japanese and instituted iron discipline to keep pilots from violating his rules or edicts. Like Eddie Rickenbacker in WW I, he preached not to attack unless you had the advantage. With the P-40, it was "one

high-speed pass and haul ass" using superior speed throughout.

The results are history. The AVG destroyed 297 confirmed Japanese aircraft with 150 more probably destroyed for a loss of only 12 P-40s and 10 pilots in combat.

While the AVG was clobbering the Japanese, the rest of the world was being clobbered by the Japanese. The U.S. Army Air Force and Navy was finding out how to fight the Japanese the hard way.

As you can see the argument between maneuverability, rate of climb, and speed continued. It was the pilot who could best use his fighter's strengths against the other guy's weaknesses who still won, but from now on he also had to worry about his opponent's wingmen as well. However, if the attacking pilot and his wingman were properly trained and motivated, they would get the opposing leader and all of his wingmen!

NOTES

1. Paul O'Neil, *The Epic of Flight:* "Barnstormers & Speed Kings", Time-Life Books, Alexandria, Va., 1981, p. 6.
2. Brendan Gallagher, *Illustrated History Of Aircraft*, Longmeadow Press, Norwalk Cn., 1977, p. 47.
3. Enzio Angelucci, with Peter Bowers, *The American Fighter*, Orion Books, New York, 1985, pp. 9-11.
4. O'Neil, Op. Cit.; pp. 109-139.
5. Ibid.
6. Karl Ries, *The Legion Condor:* Appendix F, Ballantine Books, New York, 1992, pp. 254-263.
7. O'Neil, op. cit.; pp. 84-99.
8. Arch Whitehouse, *The Military Airplane*, Doubleday & Co., New York, 1971, pp. 164.
9. Lloyd S. Jones, *U.S. Bombers*, Aero Publishers, Fallbrook, Ca., 1980, pp. 28-32.
10. Enzo Angelucci, *The Rand McNally Encyclopedia of Military Aircraft 1914-1980:* Plates, 79, 83, 84, 85, 87, & 92; The Military Press, Milano, Italy, 1983, pp. 180-193.
11. William Green, Gordon Swanborough, *The Complete Book Of Fighters*, Smithmark Publishers, New York, 1994, pp. 407-408, 425- 426.
12. Ibid.
13. Whitehouse, Op. Cit.; pp. 182-206.
14. Dewitt S.Copp, *A Few Great Captains*, Doubleday & Company, Garden City, New York; 1980; pp. 454-457; In October 1938, the Acting Secretary Of War, Louie Johnson was summoned by President Roosevelt to increase the production of U.S. aircraft. Roosevelt suggested 31,000 planes in the next two years, and 20,000/year thereafter. The budget had been expanding from almost nothing to 2,300 planes per year by 1939. In November, Roosevelt announced he would ask Congress for funds for 20,000 planes per year with a production capacity of 24,000. This meant there was much work to do. General Hap Arnold called it the "Air Force Magna Carta". He tasked his staff to build a fully manned, 10,000 plane air force by the end of 1940.
15. Ibid.
16. Ray Wagner, *American Combat PLanes*, Doubleday & Co., Garden City, New York, 1968, p.201.
17. Coop; Ibid.
18. Angelucci, Op. Cit.; pp 56.
19. Angelucci, Op. Cit.; p. 52.
20. Oliver E. Allen, *The Epic Of Flight:* "The Airline Builders", Time-Life Books, Alexanderia, Va., 1981, p. 116.
21. Angelucci, Op. Cit.; p.60.
22. Allen, Ibid.
23. Ibid.
24. Angelucci, Op. Cit.; pp. 348.
25. Allen, Op. Cit.; pp. 116-117.
26. Ibid.
27. Allen, Op. Cit.; pp. 117-118,
28. Ibid.
29. Ibid.
30. Jones, Op. Cit.; pp.68-82.
31. Wagner, Ibid.
32. Whitehouse, Op. Cit.; pp. 185-186.
33. Whitehouse, Op. Cit.; pp. 192-193. Batchelor, Op. Cit.; p.79.
34. Wagner, Op. Cit.; 389-394.
35. Ibid.
36. Batchelor, Op. Cit.; p. 54.
37. Deighton, Op. Cit.: Table 1; pp. 86.
38. Allen, Op. Cit.; p.116.
39. 100 Octane fuels were an American oil refining development made possible with the discovery of tetra-ethyl lead. The Germans did not produce 100 Octane fuels during WW II and this power boon was only available to American aircraft and the Allies.
40. H.H. Hurt, *Aerodynamics For Flying Personnel;* University Of Southern California Aerospace Safety Division, Los Angeles, 1964, pp.141-143.
41. Harold Mansfield, *Vision, A Saga Of The Sky*, Duell, Sloan, & Pearce, 1956, pp. 134-135.
42. Alfred Price, *Fighter Aircraft, Arms And Armour, London, 1990, pp. 87-88.*
43. *Whitehouse, Op. Cit.; p. 192,*
44. Martin Caidin, *Forked-Tailed Devil:* The P-38, Ballantine Books, New York, 1971, p. 122-124.

45. Whitehouse, Op. Cit.; pp.206-207.
46. Sims, Op. Cit.; p. 160.
47. Hurt, Op. Cit.; pp. 145-149.
48. Ibid.
49. Allen, Op. Cit.; p. 126. The Douglas DC-1 was equipped with constant-speed props in 1933.
50. Ibid.
51. Len Deighton, *Fighter The True Story Of The Battle Of Britain*, Johnathan Cape, London, 1977, pp. 109-110.
52. Whitehouse, Op. Cit.; p. 173.
53. Deighton, Op. Cit.; p. 107.
54. Deighton, Op. Cit.; p. 97.
55. Deighton, Op. Cit.; p. 108.
56. William Green, *Japanese Army Fighters*, Arco Publishing Co., New York, 1978, p. 15.
57. William Green, *The Complete Book Of Fighters*, Op. Cit.; pp. 205-206.
58. Green, *Complete Book Of Fighters*, pp. 134-136.
59. Op Cit; pp. 473-475.
60. Deighton, Op. Cit.; 97-98.
61. Whitehouse, Op. Cit. pp. 172-175.
62. Wagner, Op. Cit.; p. 201.
63. Deighton, Op. Cit.; pp. 103-105.
64. Deighton, Op. Cit.; p. 98.
65. Deighton, Op. Cit.; pp. 105-108.
66. Martin Caidin, *Fork-Tailed Devil:* The P-38, Ballentine Books, New York, 1971, p. 31.
67. The author's basic pilot instructor used to comment on the similarity. Tony Noonan had flown the P-38 in North Africa and Italy during WW II. According to Tony, the noses of the F-80 and T-33 were nearly identical with the P-38. Since they were all manufactured by Lockheed – could be.
68. Wagner, Op. Cit.; p. 219.
69. Green, *The Complete Book Of Fighters;* reference individual designs.
70. Mike Spick, *Fighter Pilot Tactics*, Stein and Day Publishers, New York, 1983, p. 36; 1922 RAF Training Manual.
71. Ibid.
72. Ibid.
73. Batchelor, Op. Cit.; pp. 134-135.
74. Spick, Op Cit.; p. 41.
75. Ibid.
76. Ries, Op. Cit.; Appendix L, p.275.
77. Spick, Op. Cit.; pp. 43-44.
78. Ibid.
79. Ibid.
80. Deighton, Op. Cit.; p. 218.
81. Spick, Ibid.
82. Green and Swanborough, *Japanese Army Fighters*, Op. Cit.; pp. 2-15.
83. Duane Schultz, *The Maverick War, Chennault and The Flying Tigers*, St. Martin's Press, New York, 1987, pp. 1-15.
84. Op. Cit.; pp. 113-116.
85. Ibid.
86. Op. Cit.; p. 282.

Above: *Designed in the mid 1930's the Hawker Hurricane was the most numerous fighter used by the British in the Battle of Britian in 1940. The Spitfire was faster, the Hurricane more maneuverable and more rugged. It used the same Merlin engine and eight guns as the Spit and was about 20 mph slower. The "Hurris" usually attacked the bombers while Spits took care of escorting fighters.*

CHAPTER 3

THE AIR WARS BETWEEN WORLD WARS
THE SPANISH CIVIL WAR

THE COMBATANTS

The Spanish Civil War is sometimes characterized today as a war between Fascism (Nationalist) and Communism (Republican). To any serious student of the war this is a vast oversimplification. However, the Spanish Civil War is interesting to students of Fighter Air War because it was the testing ground for the best fighter aircraft of the day. The Spanish Republican Government was supported by the Soviet Union with Polikarpov I-15 biplane and I-16 monoplane fighters.[1] The Spanish Nationalists were supported by the Germans and Italians. General Franco, the Nationalist Leader, was transported from Spanish Morocco back to Spain by the German Luftwaffe. The Germans supported the Spanish with air power that included the Heinkel He-51 biplane, He-112 monoplane and Messerschmitt Me-109 monoplane fighters. The Italians supported the Nationalists with Fiat CR-32 biplane and G-50 monoplane fighters.[2]

The British, French, Americans, and Japanese avoided commitment and, honoring an international embargo, did not officially sell aircraft to either side. Incidentally, the Nationalists consistently and erroneously listed opposing fighters as American Curtisses. These were in fact Russian I-15 models powered by Curtiss-Wright radial engines. In mitigation, the I-15 does slightly resemble the Curtiss Hawk fighters. The I-16, also powered by Wright engines, was often listed by the Nationalists as a "Boe-

ing" although the Boeing P-26 had a fixed landing gear and looks entirely different. Aircraft recognition in Spain was poor particularly for the Nationalists.[3] The Germans in Spain seemed to refuse to believe that the Russians could build modern aircraft.

If you are among those who thought Russians were primitive peasants prior to WW II (as the Germans evidently believed at the time), the Polikarpov I-16 is a bit of a shock. It was the first operational monoplane fighter with a retractable landing gear in the world and for a time the best fighter in the world. It pioneered pilot seat armor, and arrived in Spain equipped with four ShKAS 7.62 machine guns. By 1938 these four guns had become two machine guns and two ShVAK 20 MM cannon (the cannon were wing mounted and the MGs synchronized to fire through the prop). In 1939 when the Me-109E and Spitfire had caught the I-16 in performance, this armament package with its terrific firing rates delivered twice the weight of fire of the Me-109 and nearly three times the weight of the Spitfire. This projectile weight was not duplicated by contemporaries until 1941.[4] As we will see later in Manchuria, this armament was devastating to the lightly constructed Japanese fighters it opposed. Hardly primitive design!

THE AIRCRAFT

Among the biplanes, the I-15 outperformed the He 51 and was faster than the CR-32, but the CR-32 could outmaneuver it. The CR-32 was considered the most maneuverable fighter of the war.[5] When the I-16, showed up, it

was superior to all of the biplanes.[6] The short-coupled I-16 "Rata" as the Nationalists called it, was very unstable at slow speeds and a handful to land, but as the first modern monoplane fighter in combat, it was dominant.[7]

The first Me-109 appeared at Nationalist airfields in December 1936 for operational tests. The Me-109B-2 equipped with a VDM variable pitch prop went into squadron service in April 1937. This fighter outperformed the I-16, was faster, and could dive away from the I-16s. The dominance of the I-16 was ended by the Me-109 although the I-16 was still a formidable opponent. Progressive development continued during the war with Me-109Cs and 109Ds being delivered. The Me-109E of Battle of Britain fame was too late to affect events in Spain.[8]

The He-112 is something of a mystery. Seventeen were introduced in Spain in 1938. Luftwaffe pilots were enthusiastic before the fact, but no details were recorded as to actual performance and they were quickly assigned to a Spanish squadron. They are dismissed with the terse comment: "No discernible advantages over the Me-109".[9]

Details of the Republican forces are sketchy today. The Republican Air Forces were a "kluge" of various aircraft until Russia provided extensive support in the form of hundreds of fighters and bombers. The Soviets also provided licenses for manufacture of their aircraft in Spain. The Russians continued to improve the I-15s and I-16s all during the war, incorporating better weapons, aerodynamic improvements, and stronger, better engines.[10] The Nationalists called these "Super Chatos" (I-15s), and "Super Ratas"

(I-16s) and were sensitive to their improving performance. This was the impetus for the introduction of the Me-109 and He-112 by the Germans.[11]

The Italians also produced a low-wing, all-metal monoplane, the Fiat

G50. Twelve were sent to Spain for evaluation in 1939. However, like the He-112, little information is now available. From performance figures available today, the G50 did not outperform the I-16 and with only two 12.7 MM

machine guns did not outgun it either.[12]

THE FIGHTER PILOTS

The fighter pilots of this war were a collage of international flyers, but two patterns are apparent:

The leading Nationalists Aces were primarily Spanish:

- Morato - 40 kills

- Diaz-Benjuma - 24

- Sagastizabal - 21

The leading Republican aces were foreigners.

- Goronov - 22 (Bulgaria)

- Serov - 13 (Russia)

- Suprun - 12 (Russia).

Republican fighter pilots were from Spain, France, Yugoslavia, Russia, U.S., and Bulgaria, to name just a few. Some were highly paid soldiers of fortune; others were dedicated Communists ready to fly for nothing but subsistence. The Nationalists on the other hand were professional airmen representing their countries even though a facade of volunteerism was maintained by Italy and Germany. The fighting was bitter. Civil Wars are sanguinary, and

_____

Left: *When the Me-109 appeared in Spain it revolutionized fighter air war and until it encountered the Spitfire, dominated the air for Germany. It remained in production throughout the war.*

heartless, and this civil war between the ideologies of Fascism and Bolshevism was worst than most.[13] The Spanish seem to have an affinity for passion whether love or hate.

Most of the battles in the air reflected this blood-thirsty mode, but one duel was unusual. It occurred between Bruno Mussolini and a superb American pilot, Captain Derek D. Dickinson. Dickinson was the squadron commander of the Escadrilla Alas Rosa ("the Red Wings") and they had tangled with Mussolini's Italians many times. Bruno Mussolini was the son of the Fascist Dictator and ruler of Italy, Benito Mussolini. Bruno issued a personal challenge to any pilot in the Red Wings to meet him in personal, one on one combat. [14]

After a month's dialogue of insults, Dickinson and Mussolini finally agreed to meet in late August 1937 in the best chivalric manner of WW I. It was agreed that anytime one of the opponents had enough he would yield by throwing a gauntlet over the side with a large white scarf attached (three feet wide and six feet long). They would pass head-on at 15,000 feet and the fight was on. One observation aircraft per side would circle the combat area one thousand feet above the duel. No other aircraft were permitted near the combat.[15]

Dickinson was in a I-16 Mosca with four guns firing through the prop and two added wing guns. Mussolini was flying a souped-up new Fiat with a powerful Hispano-Suiza engine. They fought for 22 minutes, an eternity in air-to-air combat. Dickinson was forced down behind his armored plated seat numerous times as Mussolini riddled his ship on several passes. The American used his

"last ditch" maneuver at least twice to re-gain the advantage. The last time he hammerhead stalled off the top of a high G loop and dropped on the Italian's tail as Mussolini overshot. As Dickinson was about to fire, a gauntlet flew from the Fiat's cockpit. Dickinson and Mussolini dipped their wings in salute with Dickinson the acknowledged winner.[16] All this amid one of the bloodiest civil wars in history, where cruelty, and amorality ran amuck. It is understandable that many think this fight was fiction.

SUMMARY

The Spanish Civil War provided a preview of the fighter air war to come in 1939 when WW II began. It precipitated the end of the biplane fighter in the West as the Sino-Japanese War did in the East. The "fluid-four" (Rotte/Schwarm) formation developed by the Legion Condor still dominates fighter tactics in the missile age nearly 60 years later. The Germans rediscovered close-support during this war and developed procedures of working with the Army by radio that were revolutionary and that contributed heavily to the success of Blitzkrieg in WW II. Actually, this does not seem to be that much progress after three years of bloody war.

In fact, when looked at in the broader air power picture, Spain misled the Germans. First their leadership failed to perceive the all-out importance of fighters and their role in air superiority and air supremacy as it related to the strategic mission both in offense and defense. This failure would cost them drastically, first in the Battle of Britain (offensive) and

later in the Air War over Germany in 1943-1944 (defensive). Instead, the Air leadership was hypnotized by ground support and tactical bombing.[17] Again although it was clear to everyone who fought there, the importance of fighters in the overall scheme of air power was missed again. To misunderstand the fighter's role in air war was a trend that started in WW I and as we will see continues until today.

In Spain, we can see the same pattern of fighter development affecting the air war as we did in WW I. One side would develop a superior fighter and dominate the air. First, the I-16 Moska (Rata to the Nationalists) dominated for the Republicans, then the Me-109 turned the battle around for the Nationalists, exactly as we saw in WW I with the Eindecker and Nieuports.

The maneuverable Fiat CR-32, supplied in great numbers, held its own throughout the war.[18] In fact, all of the leading Nationalist aces flew the CR-32. The parallel here would appear to be to the highly maneuverable triplanes in WW I. The success of the CR-32 may have been due to the poor air discipline of the Republicans. The faster, more heavily armed I-16 had to slow down and dogfight to be vulnerable to this slower, more maneuverable foe.

Unfortunately, no Republican air doctrine, if it existed at all, survives today. This success also misled the Italians doctrinally. Their primary fighter was still a biplane in the 1940s and her pilots resisted enclosed cockpits even when monoplanes superseded their beloved biplanes.

The bottomline is the Spanish Civil War gave the Germans and Italians valuable combat experience that proved useful in WW II several years later. However, it did not supply any guarantees and both countries made major mistakes in the employment of air power largely because of the experience and the erroneous conclusions their leaders drew from their Spanish experience.

THE SINO-JAPANESE WAR

THE AIRCRAFT

The Japanese invaded China from Manchuria in July of 1937. This war became a proving ground for many contemporary fighters and unlike Spain there was no embargo by the western democracies. Japanese Army and Navy used this war as a proving ground like the Germans and Italians used Spain. The Japanese Mitsubishi Type 96 and Nakajima Type 97 monoplanes dominated the air war. The Chinese countered with American, Curtiss Hawk III/IVs biplanes, Hawk 75As monoplanes(fixed gear P-36s), Russian I-153 biplanes (I-15s with retractable gear) and I-16 monoplanes. Captain Claire Chennault USA (RET) was hired as a Chinese adviser and he conducted an aerial "guerilla war" against the Japanese. He taught his Chinese pilots to only attack when they had overwhelming superiority and to disappear when outnumbered or at a disadvantage. He organized a highly efficient early-warning network which made these tactics possible.[19]

The Japanese never were able to totally eliminate this small Chinese Air Force. The worst moments came in 1940 when the Mitsubishi Type 0 was introduced. This was the famous "Japanese Navy "Zero" fighter of WW II fame and was the best carrier-qualified fighter in the world at this time. It could outperform anything in the air over China. Armed with two fuselage mounted synchronized 7.62 MM machine guns and two 20 MM wing cannon, the Zero had a combat radius of over 500 miles on internal fuel.[20] There were also provisions for an external droppable fuel tank that extended this range even further. It was a formidable aircraft and would decimate Allied air forces in the early years of WW II.

This range capability is interesting and was a startling surprise to Allied forces in WW II. Because of Chinese withdrawals beyond fighter range, Japanese bombers were subject to Chinese fighter attack. The Japanese needed to escort their bombers to stop Chinese fighters from decimating their bomber formations. To meet this requirement, the Japanese Type 96 was fitted with a 200 gallon external drop tank in 1938, and this capability was a design feature of Japanese fighters thereafter. As we shall see, this provision for the Me-109 in the Battle of Britain might have turned the tide for the Germans. To everyone but Chennault, the flexibility and superiority of Japanese fighter design was a rude shock in WW II.

The fixed gear Type 96 and 97 aircraft were able to more than hold their own with the Russian I-153, I-16 and American Hawk fighters. The Japanese fighters were extremely nimble and maneuverable. The untrained Chinese and International pilots had to be taught and trained not to dogfight with these exceedingly maneuverable aircraft. This was extremely difficult to do in an undisciplined, multi-lingual force. Chennault knew what to do, but he needed unlimited authority, better equipment, and a homogeneous force and unit to get it done. Therein lies the genesis of the American Volunteer Group, The Flying Tigers. The dictum, "Do not dogfight with the Japanese" was born over China.

THE FIGHTER PILOTS

The Japanese fighter pilots were a confident, homogeneous, well-trained, and disciplined group. They believed their aircraft were the lightest, most maneuverable in the world (with some reason). Agility and maneuverability were almost a fetish with these pilots. The oldest Japanese martial arts tradition, "Kendo", values dexterity and agility in fighting above most other attributes.[21] Raised in this tradition, the desire to dogfight becomes understandable as does the design of fighters to express this tradition. Intense and dedicated, these pilots became experienced and deadly opponents over China. The average time of Japanese carrier pilots at Pearl Harbor was 800 hours and almost all had seen combat. The U.S. Navy, in contrast, averaged half of that with no combat experience.[22]

The Chinese Air Force was a conglomeration of pilots similar to those soldiers of fortune who fought for Republican Spain. Multi-lingual, with different training and experience, and indifferently led, they suffered in training and discipline when compared with the Japanese. Desperate to combat the Japanese in the air, the Chinese first tried an American air mission, and then an Italian air group. Both were failures. These were followed by the successful Chennault.[23]

SUMMARY

The Sino-Japanese War was like the Spanish Civil War: a preview of WW II. It dramatically pointed out what was going to happen between the Japanese and the Allies during the early days of the coming world war. But because the Sino-Japanese war was poorly documented in the west, and because of western technological arrogance, everyone in the West was astonished at the excellence and facility of Japanese fighters at the beginning of WW II. In many cases, Chennault had warned the Americans, but he was not believed.

The Japanese were one of the first groups to realize that bombers needed fighter escort and the need, therefore, for a long range in their fighters. They were among the first to develop external drop tanks and the first to fit that capability to all of their fighters. By 1940, Type Zero fighters were flying escort missions of over 1,200 miles round-trip. An unheard of distance for a fighter in 1940.[24]

Claire Chennault learned how to fight the Japanese fighters after studying their equipment and tactics for several years. By using his heavier, faster, better armed and armored fighters, fluid two and four formations, and the dive and zoom attack, he was able to overcome the more maneuverable Japanese. He then proved these tactics with the AVG.

Until these lightweight fighters were replaced, these procedures and tactics were effective against Japanese fighters throughout WW II.

THE RUSSO-JAPANESE WAR

THE COMBATANTS

On May 11, 1939, a Soviet cavalry force crossed the Mongolian border into Manchuria and initiated what the Japanese called the "Nomonhan Incident".[25] This incident triggered some sizeable air battles over Manchuria. On 22 May the first fighter conflict took place between Japanese and Russian fighters. During the month of May, there were three engagements between Japanese Type 97s and Russian I-16s and I-152s. Although always outnumbered, the Japanese claimed 54 victories with minimal losses to themselves during these fights.[26]

Before evaluating these claims, it must be recognized that most of these Russian pilots were embryo fighter pilots with 50 to 90 flying hours operationally. The Japanese were seasoned combat trained China veterans.[27]

On June 22, the biggest air battle since WW I took place. From then on the fighting escalated. By July, the Japanese were showing signs of battle fatigue. Always outnumbered, they were now faced with a more disciplined Russian fighter force that refused to dogfight. Disturbingly, the I-16s they were now fighting were armed with two ShVAK 20 MM cannon as well as two machine guns. These aircraft were heavily armored, had self-sealing tanks, and were difficult to knock-down with two rifle cal-

iber machine guns. The I-16s attacked in formation using superior speed and dives and zooms. Japanese losses escalated. The last battles of the "Nomonhan Incident" took place in September.[28]

SUMMARY

This "Incident"/War reflected the coming world war fighter experience almost exactly. The inexperienced Russian pilots were decimated exactly as the inexperienced Allies were later on. The agile Japanese fighters, flown by combat veterans, appeared unbeatable to both. When cooler heads gained experience, however, tactics evolved to defeat the Japanese. It is apparent that Chennault for one was doing his homework. The Russians also may have profited from the experience of attacking the maneuverable Italian CR-32s earlier in Spain. Their tactics adjusted admirably over Manchuria.

NOTES

1. William Green & Gordon Swanborough, *The Complete Book Of Fighters*, Smithmark Publishers, New York, 1994, pp.476-478.
2. Green & Swanborough, Op. Cit.; pp 205-208.
3. Carl Ries & Hans Ring, *Legion Condor*, Schiffer Publishing, West Chester, Pa., 1992, p. 37.
4. Alfred Price, *Fighter Aircraft*, Arms and Armour Press,   London, 1990, pp. 13-14.
5. Ries, Op. Cit.; p. 28.
6. Op. Cit.; p. 35.
7. Price, Ibid.
8. Ries, Op. Cit.; pp. 261-262.
9. Ries, Op. Cit.; p. 258.
10. Green and Swanborough, Ibid.
11. Ries, Op. Cit.; pp.263.
12. Green and Swanborough, Op. Cit.; pp. 209-210.
13. Ries, Op. Cit.; Appendices G & H; p. 264.
14. Martin Caidin, *The Ragged Rugged Warriors*, Ballantine Books, New York, 1966, pp. 27-30.
15. Ibid.
16. Ibid.
17. Ries, Op. Cit.; pp. 234-235.
18. John Batchelor & Brian Cooper, *Fighter*, Charles Scribner & Sons, New York, 1973, pp. 74-75.
19. Batchelor, Op. Cit.; p. 79.
20. Ibid.
21. Green, William & Swanborough, Gordon; *Japanese Army Fighters;* Arco Publishing; New York; 1978; pp. 3.
22. Walter J. Boyne, *Clash Of Wings*, Simon & Schuster, New York, 1994, p. 104.
23. Martin Caidin, *The Ragged,Rugged Warriors*, pp. 51-53.
24. Op. Cit.; p. 104.
25. Green, *Japanese Army Fighters*, Op. Cit.; p. 10.
26. Ibid.
27. Ibid.
28. Ibid.

Above: *The Grumman F4F-3
"Wildcat" became the U.S. Navy's
standard shipboard fighter in 1940.
Slower and much less maneuverable
than the Mitsubishi AGM5 "Zero", it
was much more rugged and could
"dive and zoom" with its nimble
opponent. Engine: 1200 hp Twin
Wasp Pratt Whitney. Armed with
four .50 (12.7mm) caliber guns. Top
Speed 321 mph. U.S. Navy tactics
capitalized on teamwork and the
rugged capabilities of this aircraft.*

CHAPTER 4

FIGHTER DEVELOPMENT IN WORLD WAR II

1939-1943

INTRODUCTION

The Second World War followed the first by only 21 years, but the technological advancement during that time was phenomenal. We advanced from a basically steam-powered society to one of nuclear power. The fighter advanced from a literal powered kite to a machine that sped at near sonic speeds. It became possible to win wars from the air, but the essentials of air warfare as developed in WW I remained largely the same. Unfortunately, many had forgotten what had been developed then and many had to die before those fundamentals were rediscovered. One of those forgotten essentials was the efficacy of the fighter.

A careful review of the development of air power doctrine prior to WW II runs into a surprising fact. Although all airmen everywhere bought Douhet's theories generally, there were widely differing views among air power adherents in each country and major differences among the air forces in each country.[1] For example, Germany in surveying Europe, concluded that large numbers of mid sized bombers could make any mass bombings required. The British, with the channel, Zeppelin experience in WW I, and the massed threat of German bombers growing, placed fighter defense high in their top priorities, but did not stop work on "strategic" long-ranged four engined bombers.[2] The strategic bomber dominated the upper command structure of the RAF as it did the U.S. Army Air Corps.

In France, most of the emphasis was on the army and army support. There were no international air power figures and her air forces stagnated. In the U.S., the "Strategic Air" school predominated, but the parent army kept attack aviation and smaller bombers alive. The Navy fought its own internal political "war" between the aircraft carrier advocates and the "battleship admirals". Both the Army Air Corps and Navy fought each other for the very slim budget dollars. The fighter doctrine people in the Army Air Corps and the Navy were shunted to the side. This domination was total and continued into recent times. No fighter pilot was appointed as Air Force Chief Of Staff until the 1980s.

It is no wonder that all-out fighter doctrine people like Chennault were ignored and forced to retire. They were given voice in combat but still were treated poorly by the "bomber boys" even after they had proven themselves right. Unfortunately, "Chennault could get along with no one above him except those who agreed with him."[3] And all of those above him disagreed with him, particularly General H.H. "Hap" Arnold, soon to be Chief Of Staff.

Between wars, the fighter tactics that had developed during WW I were forgotten or lost in many national air forces. In Britain, France, the U.S., and Germany, fighter tactics were based on the three ship Vee or Vic (as discussed earlier). Japan and Italy were still enamored with single-ship, dog-fighting dedicated to the maneuvering school of fighter tactics. The Spanish Civil War and the Sino-Japanese War introduced the seeds of change and re-discovery of the lessons of the first World War. Many were to die unnecessarily in early WW II while these processes were on-going.

The development of U.S. Navy fighter tactics at the beginning of WW II was very rapid and dynamic. The fighter pilots of the Navy developed their own version of the Rotte and Schwarm under the pressure of combat. The "Thach Weave" provided at least a head-on shot at the lighter, unarmored Jap Zeros attacking an accompanying F4F (wingman or leader). As many members of the AVG were former Navy and Marine flyers there may have been some cross-pollination of tactics as these Chennault-trained pilots reentered their former services. But for whatever reason very early in the war, the Navy developed tactics fighting in pairs, using "Fluid Four" spread formation, and dive and zoom tactics against the nimble Zekes, Nates, and Claudes.

Engines continued to grow in power with radials topping out at 3500 HP. This became irrelevant at the end of the war when turbine engines appeared and boosted aircraft speeds over 500 MPH well beyond the capabilities of any prop aircraft.

In aircraft design, WW II engineers were faced with a whole new ball game in aerodynamics. Modern WW II fighters could reach 500-600 MPH in dives and at these speeds airflow over control surfaces could reach the speed of sound where air molecules begin to compress and all the physics of flight change. The aerodynamics of shock waves were not well understood.

With the capability to eliminate entire cities, air superiority and air supremacy took on new meanings. Douhet's premises and Stanley Baldwin's statement "The bomber will always get through"[4] took on new meaning with the employment of the Atomic Bomb. The controversy about whether wars can be won solely from the air raged anew as Japan surrendered in the face of atomic destruction.

During WW II, the preeminence of the fighter in air warfare was reiterated again and again,[5] but as we shall see, the narrow-minded and the closed-minded are hard to convince.

PART ONE---1939-1940

EUROPE

In September 1939, the Germans invaded Poland using the same techniques and propaganda that they had used earlier to take over the Rhineland, Austria, and Czechoslovakia. After Munich, Hitler was sure that the British and French did not have the courage to act and disregarded their warnings that they would go to war to support Poland if Hitler invaded. Hitler first concluded a non-aggression pact with the Soviet Union, and then the Germans marched into Poland. Before we review the Polish campaign, let us review the best fighters of the world at the start of the war. Following our premise that air superiority often goes to the country with the qualitatively best fighters, such a review is enlightening. The following lists the best operational fighters in each country as of September 1939:

• RAF – Supermarine Spitfire, arguably the best fighter in the world. 1,100 HP. Rolls-Royce Merlin. 367 MPH, Eight .303 caliber machine guns.[6]

Below: *The P-39 was one of the beautiful aircraft of the war, but lack of a turbo-supercharger limited performance. It served well at low altitudes in the Pacific and Russia. Note the engine mounted mid-ship above the wing to improve maneuverability & provided room for a massive 37mm gun forward.*

- Royal Navy – Gloster Sea Gladiator, obsolete biplane, 830 HP Bristol Radial. 257 MPH, two synchronized .303 caliber & two .303 caliber wing-mounted guns.

- Germany – Messerschmitt Me-109E, about equal to the Spitfire, 1,100 HP. Daimler Benz DB 601A, direct fuel injection and a variable-speed supercharger. 354 MPH, two 20 MM wing cannon, two 7.9 MM machine guns.[7]

- France – Curtiss Hawk 75A, export version of the American P-36, 1,200 HP Pratt-Whitney radial. Constant-speed prop. 311 MPH, one .50 caliber, one .30 caliber synchronized and two .30 caliber wing-mounted machine guns.[8]

- U.S. Army – Curtiss P-36, 1,200 HP Pratt-Whitney radial. Constant-speed prop. 311 MPH, one .50 caliber, one .30 caliber synchronized, and two .30 caliber wing-mounted machine guns.[9]

- U.S. Navy – Grumman F3F-4, biplane, 950 HP Wright Cyclone radial. 256 MPH, one .30 caliber, and one .50 caliber synchronized machine gun.[10]

- Russia – Polikarpov I-16, 800 HP radial, 304 MPH, two 7.62 MM synchronized machine guns, two 20 MM wing cannon.[11]

- Japanese Army – Nakajima Type 97 (NATE), Nakajima Ktobuki 780 HP radial, 292 MPH, two synchronized 7.62 MM machine guns.[12]

- Japanese Navy – Mitsubishi Type 96 (CLAUDE), 785 HP Kotobuki Radial. 270 MPH, two synchronized 7.62 MM machine guns.[13]

- Italy – Macchi C. 200 Seatta, Fiat 870 HP Radial, 312 MPH, Twin synchronized 12.7 MM machine guns.[14]

* Poland – PLZ-11c, Bristol Mercury (Polish Built) 540 HP Radial, 242 MPH, Four wing-mounted 7.62 MM machine guns.[15]

Note that all of the countries were using low-wing all-metal monoplanes with retractable landing gears as their best fighters except the U.S. Navy, the Royal Navy, Japan, and Poland.

The U.S. Navy F3F-4 was a biplane with retractable landing gear while the Royal Navy Sea Gladiator was a biplane with fixed landing gear. The biplane with its slow landing speeds and rugged structure still appealed to the carrier navies, but note that the "backward" Japanese were ahead of the world. The CLAUDE was the first monoplane in the world to carrier qualify.[16]

The Mitsubishi Type 96 (CLAUDE) and Nakajima Type 97 (NATE) both had fixed gear with "pants" (aerodynamic wheel covers used to reduce drag). With fixed gear, they performed nearly as well as many of these fighters that had retractable landing gears. And Japan had a series of low-wing fighters with retractable gear in final development. The Mitsubishi Type 0, (ZEKE) the famous "Zero", went into operational service in 1940 and the Army OSCAR and HAMP followed.[17][18]

The Polish PLZ-11c was a high gull-wing monoplane capable of only 240 MPH, and although very maneuverable, obsolete. Even so, these early 1933-35 type fighters were successful against German bombers. But when up against Me-109Es and Me-110s they were outclassed, outgunned, and outnumbered.[19]

The Spitfire was the fastest of these fighters below 20,000 feet. Above 20,000 feet, the Me-109 was the fastest. The Russian I-16 Moska with seat back armor and two fast-firing 20 MM guns and two 7.62 MM machine guns was the best armed and armored. The Nakajima Nate was the most maneuverable of the lot. The Mitsubishi Claude was equipped with a 46 imperial gallon external disposable tank. Above 400 MPH, the P-36/Hawk 75A could outmaneuver all of them with its balanced, high mechanical advantage controls (the trick was to fight at 400 MPH or above, in a 250-300 MPH fighter). The Me-109 Daimler-Benz engine had the advantage of an automatically variable supercharger and direct fuel injection. German pilots could push straight over into a dive using negative Gs and climb back to altitude without doing anything to the engine supercharger as it automatically adjusted to increasing or decreasing outside air density. P-36/Hawk 75A pilots could set a constant RPM on their propeller and could be sure it would automatically hold that RPM regardless of throttle position. The Me-109 pilots had to adjust the prop pitch everytime they moved the throttle. When the war started, the Spitfire was losing performance because it had a fixed pitch propeller. This affected time to climb, maximum power and particularly, performance at high altitude. When in combat where every ounce of thrust was critical, these losses were serious and a two-speed variable prop was retrofitted. The Spitfire's service ceiling was immediately raised by 7,000 feet.[20]

It is apparent that no one fighter had a monopoly on strengths or weaknesses, and in airplanes anywhere near equal, it was the pilots, their training, experience and tactics that made the difference. The pilots who won were able to capitalize on their ships' strengths and take advantage of their opponents' weaknesses. This required the development of air combat doctrine, study, hard training, and preparation. You had to know your own weapon system's strengths and weaknesses (this included your own strengths and weaknesses as a pilot for you were a key part of the overall system) and the strengths and weaknesses of your opponents.[21]

There were two other important factors that were crucial in air war in 1939:

The first was fighter reliability. Reliability needed to be part of the design itself. If the engine would not run, or the systems did not perform as designed, the fighter was less effective and sometimes worthless. Sometimes these design faults could be fixed (e.g., P-38 turbochargers). Others could not (e.g, Me-410s).

The second factor contributed to the first. As fighter aircraft were complex, "state-of-the-art" machines, maintenance and supply were also critical to success. The men who maintained the aircraft were almost as important as the men who flew the fighters. Given fighter aircraft equal to those of your opponent, and equality in pilot abilities, the most reliable and best serviced aircraft usually prevailed. In studying military history, this fact is sometimes

overlooked. However, anyone reviewing U.S. performance in the air warfare of WW II would really be remiss to overlook this factor. In the end, it was U.S.'s overall logistical superiority in production and maintenance that was largely responsible for our final success.[22]

But when WW II began, we were not ready for war. During the first two years we were really scrambling to catch up in the air. There were some heart-breaking failures. For example, in Java in 1941, when we were desperate for dive-bombers, 52 A-24s arrived from the U.S. without firing mechanisms for their guns. They were next to worthless for combat.[23] It is again important to remember that it is the weapons

Right: *At the beginning of the war, the Curtiss P-40 was the best operational USAAC fighter. Developed from the P-36, it put on pounds due to the liquid cooled engine but gained significant speed from lower frontal area (parasitic drag). The P-40 had to dive and zoom with the more maneuverable Japanese Zero. At high speeds it could out-roll the Zero. With altitude available it could always dive away from its lighter adversary.*

system that prevails and maintenance and supply are key elements of that system.

The logistics of air war, in their own way, are as complex as combat tactics and had to be learned the hard way in actual wartime operations. By 1939, the Germans and Japanese had the advantage of combat experience. Their equipment, supply, and maintenance systems had been tested under actual combat conditions in Spain and China from 1936-1939. The Germans, in particular, showed great capability to move forces throughout Europe with celerity and efficiency all during the war. Their ability to deploy and operate at long distances from their home bases was aptly demonstrated as early as 1940 in Norway.[24] The British, as we shall see, had more difficulty with this important capability.

The Japanese, after fighting in China, struck with effective air power all over the Pacific in 1941. Their operations were conducted over very long distances, but were very effective and they prevailed everywhere with ease. Interestingly, the Japanese used long-ranged, land-based aircraft as part of the Imperial Japanese Navy.[25]

Another more obvious factor was numerical superiority. Both the Germans and Japanese had this advantage at the beginning of their respective wars during WW II. As we have seen, fighter technical superiority can overcome numerical superiority to some extent (the Eindeckers and Albatroses were always outnumbered during WW I)

but it becomes no contest when the numbers are overwhelming and technical quality is near-even or better for the majority. Consider WW I in 1918. The German Fokker D VII was as good as, or perhaps better than, the Camels, SE-5As, and Spads. But there were too many of them and not enough D VIIs.[26]

In modern war, good command and control can offset numerical superiority by allowing the side with better command and control to concentrate its fighters where they are most needed. Examples are the RAF with their RADAR/Operations Centers during the Battle of Britain[27] and the AVG with the Chinese early warning net.[28]

Germany and Japan had the edge in production build up for WW II. As they were the aggressors, they were prepared for action when they struck. Hitler's propaganda had the world believing that the Luftwaffe was the strongest Air Force in the world.[29] It probably was, but as we have seen, air superiority/supremacy is fragile and must be nurtured. A technological breakthrough can change the picture overnight and fighter weapons

49

systems were the key.

But neither Japan nor Germany had learned two vital lessons of air power:

1) Air power (fighter) operations are continuous and losses are at best constant. Air superiority must be fought for steadily; there is no facility for let-up. Training and production must be continuous to replace losses with no interruption possible until the conflict is over.[30]

This dictates that not only production, but research and development must continue unabated as no one can predict how long the conflict will last.

2) Air Forces must have large standing reserves to meet operational surges and unexpected losses. Even the most powerful and numerous air forces encounter difficult problems after unforseen setbacks. Only by building reserves can they accommodate to these and continue to prevail.[31]

This is reflected in the production figures of aircraft found in the mobilization plans for Britain, Germany, and France. The British became alarmed after Munich in 1938 and launched an all-out effort to build up their air forces. The following table representing these mobilization plans is interesting:

• Britain – produce 2,000-3,000 planes of all types per month within 18 months. Britain revised this plan to 1,700 planes/month in September 1940 based on experience.

• France – produce 780 per month by the seventh month. In actuality France built 2,100 aircraft in the first five months of 1940.*

* It should be noted that both the British and French were counting on U.S. production facilities to assist in their urgent catchup. Britain alone ordered 14,000 aircraft in the U.S. by August 1940.[32]

The Germans on the other hand perhaps misled by the ease of the campaigns in Spain and Poland never planned for more than 1200 planes a month and actually scaled down to about 1000/month until 1942. British production surpassed the Germans by mid-1939.[33]

THE POLISH INVASION

When Germany invaded Poland, she committed over 200 Me-109 and Me-110 fighters to the attack. Poland had about half that number of fighters – all PLZ-7s & PLZ-11s. The PLZs were highly maneuverable, but hopelessly obsolete. The Polish government built up their air force in the early 1930s, but unfortunately did not continue to update their forces. In relation to WW II, the Polish Air Force peaked early, roughly 1933-1935, and was in serious decline a few years later when they were invaded. Both the Me-109Es and Me-110s proved superior to the PLZs. When the speedy German fighters refused to dogfight, the polish fighters did not have a chance.[34]

Polish intelligence followed German preparations for invasion and quietly hid their fighter force on secondary fields all over Poland. When the German bombers struck Poland, they were intercepted by the PLZs which had survived the primary strikes by the Luftwaffe and were sitting cockpit alert. The PLZs were alerted by ground observers and successfully intercepted and broke up the first raids on Warsaw. But the PLZs could barely catch the German bombers and were vulnerable to the organized Luftwaffe fighters who were over 100 MPH faster. German fighters could hit and run at their leisure. In contrast to the British later, the Poles thought that the Me-110s were more formidable than the Me-109s. By 26 September, the Polish Air Force had lost over 83 per cent of its forces. The Poles fought bravely, but it was all over in about four weeks.[35]

The air superiority gained by the Luftwaffe was key in breaking up Polish ground counterattacks and was decisive again and again in demolishing Polish resistance. At the same time the Polish Air Force was prevented from attacking the invading German Wermacht. The world watched in amazement as newsreels showed the dominance of the Luftwaffe and films from Poland showed the effects of "Blitzkrieg". As a very little boy, I remember going to see a documentary covering Poland at a local movie with my Father. I can still remember a shot taken from inside a Stuka dive bombing a Polish rail yard. It was new and scary. Hitler's propaganda seemed real. As the war continued through Norway and France, the Germans seemed invincible. Most of this "invincibility" had its roots in air superiority gained by Me-109s and Me-110s.

To their everlasting shame, the French and British did practically nothing during the few weeks that Poland was overrun even though Hitler had stripped his forces in the West to the bone. From September 1, 1939 until May of 1940 followed what was called the "Phony War" or the "Sitzkrieg", where almost nothing transpired in the west. Even as unprepared as they were, an Allied attack into Germany during Hitler's Polish operations probably would have had some success as Germany's defenses were bare.[36]

One of the major results of French and British inaction might have been the bolstering of Hitler against the General Staff. They had counseled caution in attacking Poland as they were well aware of the capabilities of the French Army. But Hitler's estimates of the moral courage of French and British leaders was more accurate than the General Staff. The Allies sat and watched Poland go and Hitler endured.[37]

THE INVASION OF NORWAY AND DENMARK

In April of 1940, Germany, catching wind of a possible invasion of Norway by the Allies,[38] preemptively invaded Norway and Denmark. Never has a campaign more succinctly demonstrated the efficacy of air superiority/supremacy in war. By prevailing standards,

Hitler's invasion of Norway in the face of the British Navy was insanity. But times had changed, and with air superiority no navy could operate effectively. Occupying Denmark, Hitler gained bases for fighter operations and maintained air superiority as far north as Trondheim.[39]

Combined land, sea, and air operations by the Germans were successful everywhere in spite of the British Navy's attacks and Norwegian resistance. The Luftwaffe was decisive and British Navy losses heavy. The Allies countered with reinforcement of Narvik in far north Norway out of range of the Me-109s. Allied air defense was provided by a few Gladiators and a squadron of Hawker Hurricanes flown in off of the aircraft carrier *Glorious*.[40]

The Luftwaffe discovered very quickly that unescorted bombers were "dead ducks" over Narvik. The Hurricanes and Gladiators disrupted every raid. Unfortunately, with events reaching panic proportions in France, orders were received to evacuate the Allied fighters. Although not carrier aircraft, the Hurricanes and surviving Gladiators were landed on-board the Glorious safely only to be sunk by the German battleship Scharnhorst a few days later. Only two pilots of this group survived the sinking.[41]

The Norwegian Campaign brought Germany immense recognition and added to its growing aura of invincibility. The planning and execution of the Luftwaffe had been

impressive. The facility and power of air superiority was conclusively demonstrated again.[42] Interestingly, if the Japanese Zero had been developed in Britain, the RAF or Fleet Air Arm could have provided fighter cover from Britain for Narvik and the rest of Norway.[43] Range was, and is, a key factor in air power, particularly fighter range.

THE BATTLE FOR FRANCE

The French had one of the largest air forces in the world in 1918, but they stagnated terribly between wars and did not begin to build up to meet Hitler's threats until the mid-to-late 1930s. By that time, they were years behind and it was too late to rebuild their aircraft industry. They tried to close the interim gap with fighters and aircraft from the U.S. until they could rebuild their industries. They were nearly successful, but the "Blitzkrieg" of 1940 caught them with several excellent fighters in the prototype stage and not quite operational. The best was the Bloch 157 which German sources thought was better than the Focke-Wulf 190, but only one prototype had been built when France fell.[44]

When the Germans struck in May 1940, the L'Armee de L'Air was an obsolete conglomeration of American and French aircraft and unable to combat the Luftwaffe on anywhere near even terms. The 4,000 plus German aircraft swept the French and British from the air over France in one fell swoop. German air superiority was a fact from the outset and the German close support and interdiction in support of the Wermacht was instrumental in the "Blitzkrieg" that followed. Only the late model Dewotine and American Hawk 75As could offer any resistance at all, and they were obsolete. The British Hawker Hurricanes did little better. The Allied fighters were out-performed, outnumbered and outgunned by the Me-109s; however, the Me-110s were found to be not as maneuverable as the first-line Allied fighters.[45]

The Germans, as they had in Norway, started the campaign by invading Belgium and Holland with innovative attacks using air superiority, gliders, parachute troops, and link ups with rapidly moving armor. The attack on the formidable Belgian fort, Eben Emael, is illustrative. The fort, which was thoroughly modern, was struck by special airborne shock troops using gliders to land directly atop the fort after suppressive attacks by air. The shock troops used specially designed shaped satchel charges to literally blast the occupants to the surface and surrender. Prior to this attack, the Allies thought it would take weeks for the Germans to take this facility if it could be taken at all. Without air superiority, airborne operations are impossible.[46]

The French and British implemented "Plan D" and advanced north into the Low Countries to pre-planned positions along the Dyle River. They thus fell directly into Hitler's trap as he struck with massive air supported armor through the Ardennes (where the French believed no modern army could operate)[47] on the Allies right flank and past the left flank of the touted

Maginot Line. When the Germans crossed the Meuse river at Sedan, their armor broke out into the broad French plains and into the rear of the Allied armies. The key to the crossing of the Meuse and breakout was tactical airpower.[48]

The German fighters had by this time established air supremacy all along the front. The French, desperate to close the breakout, begged Churchill (who became Prime Minister the day the Germans invaded) to bomb the bridges across the Meuse at Sedan. The British responded immediately with an attack of 71 bombers supported as possible with Hurricane fighters. Of the 71 bombers, 40 were shot down by Messerschmitts or German AAA (Flak) with little, or no damage to the bridges.[49]

Attacking with their characteristic lightning speed, the German's ground forces drove through the French to the Channel coast. The German armor supported by the Luftwaffe simply moved too fast for the French to put together any organized response. Over 400,000 French, British, and Belgian forces were trapped against the Channel and North Sea.[50]

A prelude to the Battle of Britain arose over Dunkirk as the British and French fought to evacuate the trapped ground forces. On 24 May, Hitler and Von Runstedt halted the German Panzers in a very controversial decision. Reichmarshall Goering assured Hitler that he would destroy any attempt to evacuate the Allied forces.[51] The RAF committed major fighter resources to achieve air superiority over the

battlefield including a number of Spitfires. In nine days of evacuation, the Germans lost 176 aircraft while the British lost 106[52] and the British successfully embarked 338,226 British, French and Belgian troops.[53] It was the first time in a major battle that the Luftwaffe had lost air superiority.

After the Dunkirk evacuation, the Battle of France was a "walk in the park" for the Luftwaffe. An Armistice was signed on 22 June. Marshall Petain formally surrendered France on June 17, 1940. The Italians eager to claim some of the spoils, declared war on France on 10 June, but gained no ground at all against the more determined French. Italy gained nothing from the Armistice.[54]

The BATTLE OF BRITAIN

In 1939, the British had determined that 52 fighter squadrons were required for the defense of Britain. When the British Expeditionary Force (BEF) was formed, four fighter squadrons were sent to France along with 10 Attack Squadrons (Battles and Blenheims). As the war progressed and fighter losses mounted, Fighter Command reserves were tapped to replace the lost aircraft in France. By the middle of May, the RAF had dispatched 12-13 Hurricane squadrons to France. As the situation deteriorated the French Premier requested 10 more squadrons of fighters. When it appeared that the British Cabinet was about to grant Premier Reynaud's request, Air Marshall Dowding, the commander of British Fighter Command, asked for an audience with the cabinet where he

submitted a written letter stating that he could not guarantee Britain's safety in the coming Battle for Britain if the French request was honored.[55]

Churchill agonized about the French request and finally supported Dowding and did not send the fighters to France. Hindsight indicates it was fortunate that he supported his Air Marshall. Of the 261 Hurricanes already sent to France only 61 returned and many of those were in poor condition. The RAF squadrons were not trained in ground attack or in Army cooperation. They would have had to support French units as the BEF had evacuated. As it was, Britain lost 432 Hurricanes and Spitfires in May

and June of 1940 and the full force of the Luftwaffe, flush with victories in Spain, Poland, Norway, and now France, would hit Great Britain in July of 1940.[56] If Britain had lost the majority of 10 squadrons of trained interceptor pilots in May and June, she might not have made it. During the worst hours of the Battle of Britain, the RAF very nearly ran out of trained fighter pilots.[57]

The Luftwaffe was on a roll until the Battle of Britain. The Germans, along with Douhet, thought the best way to gain air superiority was to catch enemy aircraft on the ground and to destroy them by bombing. It had worked in all previous campaigns. However this success had led them to emphasize their

Above: *The best Carrier fighter in the world in 1941, the A6M Mitsubishi "Zeke" or Zero, could out-maneuver any aircraft below 250 mph. Extreemly light it could also outclimb and outrange contemporary fighters. It howovor woo undor armorod & armed.*

bombers and to neglect their fighter forces. They had always seen their Air Force as an important adjunct to their Army so when they gained air superiority they flew close support and interdiction for the Wermacht.[58] This worked well in Spain, Poland, Norway, and France. But in Britain, it was a whole new ballgame.

In Britain, the Luftwaffe for the first time faced a formidable foe in the air and a strategic air campaign where the air objective was a nation and its ability to make war. In addition, the British had a means to detect incoming raids and were difficult to surprise on the ground. Douhet never imagined electronic detection, where the defender could concentrate his forces against inbound raids. And the British had parity in the key to air superiority – single engine fighters!

The actual fighter aircraft strength as of July 20, 1940 favored the Luftwaffe:
• Luftwaffe – Me-109 – 809 aircraft, 656 in commission – Me-110 – 246 aircraft, 168 in commission

• RAF– Spitfires/Hurricanes – 609 aircraft, 531 in commission [59]

In contrast to what was generally thought, the British were out-producing the Germans in 1940 in single engine fighters. The British goal was to produce 400 Hurricanes and Spitfires a month while Me-109 production was only 140.[60] In July of 1940, British production was actually 496, August 476, and September 467.[61] The Germans grossly underestimated British aircraft production and proceeded in an almost leisurely fashion. In fact, the Germans were to remain on an almost peacetime production schedule for two more years[62] in spite of losing over 1,400 aircraft to all causes dur-ing May and June of 1940.[63]

In addition, the British Commonwealth training programs were running full tilt and concentrating on training fighter pilots. Again the German training programs were proceeding as if the war was over.[64]

AIRCRAFT PERFORMANCE

The book on the Spitfire versus the Me-109 was that the Spitfire was faster below 20,000 feet and could outturn the Me-109 at any altitude. The Hurricane was more maneuverable than both, but was also much slower.[65] For this reason the Spitfires usually took on the Messerschmitts while the Hurricanes went for the bombers. The Me-109s with fuel injection could outdive the British fighters by pushing straight over and accelerating away. The RAF fighters used carburetors and had to half-roll (split-S) to dive or the engines cut out. By using dive and zoom the Me-109s were often successful.

Interestingly, RAF pilots were taught that the Spitfire and Hurricane could outturn the Me-109. One must be really careful how one makes such statements. Turn radius depends on radial gravity; i.e., the centrifugal force the aircraft can produce in curved flight routinely measured in earth gravitational units or Gs (the weight of an object in unaccelerated flight equals one G). The aircraft that can pull the most "Gs" in a turn without stalling, or the wings coming off, turns the tightest. At 300 MPH, and 10,000 ft., the Me-109 had 8 Gs available before stalling, the Hurricane 7.5 Gs, and the Spitfire only 7.0 Gs. By the book, at this speed and altitude, the Me-109 outturns both the Spitfire and Hurricane. Note that

wing loading and wing span have nothing to do with ability to turn.[66]

Either the RAF pilots were willing to pull harder than the Germans, or the Germans were unwilling to pull 8 Gs (both are possible as the light Me-109E was known to have structural problems).[67] Also as available G varies with speed and altitude, it was possible that British aircraft could outturn the Messerschmitt at slower speeds or higher ones. Every RAF pilot and Luftwaffe pilot believed that the British fighters could turn inside the Me-109.[68]

The Me-110s quickly proved to be ineffectual in combat during the battle. They were unable to maneuver in

combat with Hurricanes and Spitfires and often needed Me-109 escorts themselves. However, by the book the Me-110 was faster than the Hurricane.[69] In spite of this, Hurricanes were often successful against the 110s.

The real key to the battle was the range of the Me-109 fighter. From coastal bases in France its combat radius was about 100 miles (about the center of London). It rapidly became evident that until air superiority was won over England no German aircraft except the Me-109 could survive. With the few short-ranged 109s available the Germans found it impossible to escort the level bombers, Stukas and Me-110s and fight off the concentrated Spitfires and Hurricanes.[70]

FIGHTERS TACTICS

The Germans used the "Finger Four" Rotte and Schwarm tactics they had developed in Spain and perfected over Poland, Norway and France. The British were still using the three-ship "Vic" developed in Europe between wars.[71] British fighters still in the tight Vics were very vulnerable as the close formation did not allow for look-out for attacking fighters. But the aggressiveness, tenacity, skill, and performance of the British once they broke up their formations and fought singly was a nasty surprise to the Luftwaffe, particularly surprising were the tenacious Spitfire pilots.[72] Before long, the advantage of wingmen being obvious, the RAF adopted the Finger Four from the Germans.[73] Interestingly, Claire Chennault had arrived at fighting in pairs from study of WW I tactics, particularly the ideas of Oswald Boelke.

As bomber forces began to take losses they began to clamor for more and closer escorts. The Me-100 pilots insisted that they must be free to attack the British fighters and that close escort was counterproductive. As we shall see, this same argument took place among the Americans in 1943-1944 where

Left: *For the first time the Me 109 met its match in the Supermarine Spitfire. The Me 109 had the advantage in dive and climb, but the Spit could usually turn out of trouble. The Mark V shown had clipped wings to improve rate of roll and speed, and two 20 mm Oerlikon Cannons plus four .303 mgs as armament.*

the final resolution made by General Jimmy Doolittle was to free the fighters from tight, close escort. Conversely Goering insisted on close escort. The German bomber pilots even insisted that the faster fighters not weave about the bombers, but slow-down and fly in close formation. This ridiculous order had the effect of making the close escort fighters almost as vulnerable as the bombers to the British fighters.[74]

THE BATTLE BEGINS

The Battle of Britain was fought in four phases:

1) Attacks on shipping, ports in the English Channel. This included attacks on some of the coastal radar stations.

2) First attacks on English fighter bases starting with Adler Tag (planned for August 12) and lasting about a week.

3) The "Critical Period" (as called by the British) 24 August – 6 September. All-out attacks by the Luftwaffe on southeastern RAF Fighter Command bases where sector operations and control centers were located.

4) A shift to attacks on London. 7 Sept on – attacks both day and night.[75]

It was a battle for which, Air Marshall "Stuffy" Dowding had been preparing for years. He had an integrated radar detection and command and control system in place.

This system was augmented by a ground observer corps such as used by Chennault in China (this observer corps was to play an important role in the battle). The system was tied together by extensive telephone and radio communications. The key was a series of operational plotting centers where the air battle could be visually depicted and key battle decisions made. The system was beyond the state of the art in 1940 and unique in the world at that time. Air Marshall Dowding was an aloof, diffident Englishman, as strong as he was silent. Perhaps his diffidence explains why he has never received the credit he was due for his creation. He literally saved the free world during the summer and fall of 1940.

Incomprehensibly, he was relieved of command immediately following the battle and took no further active role in the war. One can only wonder what we would know of this man had he possessed the charisma of a Nelson, a Trenchard, or a Montgomery. The truth though was that he was a fighter guy in a "bomber Mafia" just like Chennault, except that he was an Air Chief Marshal, and his battles for his fighters and defense systems seriously offended the British Air Staff who were air offensive – bomber oriented. He was dismissed without so much as a thank you.[76]

But old "Stuffy" had some powerful help. British intelligence had broken the German top codes. As many of the Luftwaffe orders and reports were broadcast by radio to and from Berlin, "Stuffy" had a big edge on the opposition courtesy of "Ultra" as the

program was called.[77]

"KANALKAMPF" -- PHASE ONE

The first phase of the battle began as the Luftwaffe attacked shipping in the Channel. This was an attempt by the Germans to draw RAF Fighter Command into combat at the same range differential as the Luftwaffe. During "Kanalkampf" as the Germans called it, the Luftwaffe moved forward to occupy coastal French bases from which they could attack England proper. The Spitfire and Hurricane had about the same range as the Me-109 so meeting over the Channel gave no advantage to either side.[78] The Ju-87 Stukas were ideal for attacking shipping with their accurate dive bombing capabilities.

Hitler's directive to begin the attack was dated July 2, 1940 and directed that the Luftwaffe:

1) Sweep the channel of Allied shipping.
2) Sweep the air of British fighters.[79]

The Germans wanted the RAF to launch their fighters out over the Channel where they could attack them from higher altitudes and with superior numbers. They were not aware that the British were painting them on radar and reading their orders. Courtesy of Ultra, the British were reading the German messages and orders in detail. Air Marshall Dowding knew the German plans, objectives, and timetables. He then could track their execution on his radars. When the Germans reported their losses and triumphs to Berlin the British copied those too.[80]

NOTE: There is an excellent lesson here for today's officer: -- In war, transmit electronically as if the enemy is receiving and understanding your transmissions. In Vietnam, for about 10 years the Russians were reading our transmissions, due to a communications spy (CWO Walker U.S. Navy) providing keys to our codes. The information was dutifully passed to the North Vietnamese. No wonder the North Vietnamese and VC anticipated our every move! Do not use the phone or radio unless absolutely necessary, change codes, and anticipate the enemy reading them. Set periodic code traps where the enemy must make unusual adjustments you can monitor, to see if your codes are being read. Failure to be vigilant can be disastrous.

With this information, Dowding refused to take the bait. He sent small numbers of fighter to help the merchant shipping being attacked in the channel, but refused to engage in major conflicts where his forces were at a disadvantage. Later, when the Germans tried fighter sweeps over southern England, he again refused to take the bait. He ordered his fighters to concentrate on the bombers and with his intelligence and operational systems he could distinguish between the bomber and fighter operations.[81] Unlike the Germans, the British had the best pre-

pared man in the world to lead Fighter Command.[82]

It was during this Channel phase that the weaknesses in the Luftwaffe started to appear. The Stukas were deadly when unopposed in the air, but when they were caught by British fighters without fighter escort they were decimated. This happened frequently as the Luftwaffe following Goering's orders escorted the Stukas closely. As a loaded Stuka cruised at only 160 MPH, this put the escorting German fighters at a real disadvantage against the nasty Spitfires that rose to meet them. The Hurricanes meanwhile waited for the Stuka to begin its pull-out after its dive-bomb attack (nothing could stay with the Ju-87 in a dive. Its terminal velocity straight down with dive brakes out was about 250 MPH and it could pull-out very close to the ground/water surface). Separated from their supporting fighters the Stukas were slaughtered by the Spitfires and Hurricanes. If the German fighters went down with the Stukas, they became separated anyway as the Me-109s and Me-110s couldn't stay with the dive-bombers in their dives and they gave up the precious height advantage they needed against aircraft that could outturn them. The Stukas were withdrawn from the battle early in Phase II.[83]

ADLER TAG -- PHASE TWO

The real battle began on 13 August when Luftflotten 2 & 3 targeted RAF Fighter Command bases. The Luftwaffe was to gain air superiority by attacking:

- Fighters in the air and on the ground

- Main harbors and wharfs

- RAF ground organizations

Specifically major strikes were planned against RAF Manston, Lypme, and Hawkinge, all southeastern coastal fighter bases.[84] Also the radar stations at Ventor, Rye, Pevensey, Dover, and Dunkirk (Kent) were attacked.[85] August 13, 1940 was called "Adler Tag" or Eagle Day by the Germans. Adler Tag had been planned for August 12, but was postponed one day due to weather. 1,475 sorties were flown by the Luftwaffe and 700 sorties by the RAF. The issue was drawn.[86]

The attacks on the radar stations were interesting. First of all, four of the attacks were made by "Jabos", or fighter-bombers, in this case, Me-110s (Pevensey, Rye, Dunkirk, Dover). Portsmouth harbor and the Ventor Radar were struck by Ju-88s. The Me-110s shut down all but Dunkirk in the east, but these were quickly repaired. Ventor was seriously hurt and was out for weeks, but the outage was disguised as another station imitated its signals (the ground observer corps was also rearranged to help fill the gap). German intelligence reported no outages after the strikes. Concluding the administration and operations facilities were safely underground, the Germans stopped attacking the radar sites as they appeared to be unprofitable targets. This was a decisive mistake.[87]

On the morning of Eagle Day the weather was bad in France and some bombers had failed to rendezvous with fighters. In one case, the mission had been canceled and the bombers did not get the word (their radios were channelized to the wrong frequencies). They continued to their targets without escort and were decimated by English fighters. This reinforced the belief that radar control still existed everywhere.[88]

As the battle progressed, the Germans realized that they were not prevailing and that changes were needed. The Ju-87 was withdrawn from action and the Me-110 was to be escorted. Younger fighter leaders were made commanders, and age restrictions placed on command positions.[89] The new rules went into effect on August 21, 1940 when the new Geschwader Kommodores got back from the meetings in Berlin.

As British Fighter Command bases were attacked, questions about RAF tactics arose. These differences continued and intensified as the battle progressed. Air Vice Marshal (AVM) Leigh-Mallory (12th Group Commander just north of London) thought that Dowding and AVM Park (11th Group Commander southeast of London where all the action was) were violating the principle of mass by not attacking by wings (massed squadrons) instead of piecemeal single squadrons. Dowding and Park maintained, with some logic, that there was insufficient time to form-up the "Big Wing" as Leigh-Mallory called it. Even with maximum warning, the big wings could not get formed up, climb to altitude, and intercept the Germans until after they had bombed the target.[90] Precious fuel was consumed in this massing of squadrons. The controversy was never resolved, but Dowding's

---

Below: *In 1942 a fast, maneuverable German fighter with a radial engine appeared that could out maneuver the Spitfire V. The Fw 190 was one of the best fighters of the war. Its rate of roll was phenomenal!*
The British countered by putting a bigger engine in the Spitfire MkV. The result, the MkIX was one of the best Spits of the War.

relief followed the battle. Someone on the air staff, or higher, may have agreed with Leigh-Mallory or have been anxious to settle an old score.

One factor in making this decision may have been the fact that the "Big Wings" usually intercepted German raids around London or north of London and much later in the Germans' mission when the Me-109s were out of gas, or nearly out, and had to go home. They thereby got the reputation of "turning the Germans back". This information may have influenced the Air Staff and possibly even Churchill.[91]

"THE CRITICAL PERIOD" -- PHASE THREE

On August 24 the Luftwaffe switched its attack further inland and unknowingly came very close to winning the battle. The attacks were against the bases where the key sector control centers for the air defense system were located. These centers were above ground and vulnerable. The Germans, with the best wireless intercept system in the world at this time, never targeted the British Command and Control Centers, but with these attacks almost wiped them out by accident.[92] Why they never attacked these centers based on wireless intercepts is not known. After the war, they said they had assumed that such cen-

ters would be fortified and underground, therefore not vulnerable. Remember, the Germans made this same assumption about the RAF radar/filter centers. These assumptions lead to some costly mistakes and was one of the factors that cost the Germans victory.

In any case, these attacks were what Dowding feared the most. Even though his attrition in aircraft was high, and his command and control facilities hurting, he still had enough aircraft and control to effectively fight. It was his aircrews that were wearing down. There were insufficient replacements and his pilots were dead tired. Many were flying 3-4 sorties a day. His rotation of aircrews to quiet sectors for rest was disrupted. Training standards were lowered, his experience levels falling and the two air forces were slugging it out over his key bases everyday. It was very close, and from 24 August to 6 September it looked as if the Luftwaffe was going to win.[93]

Aware of how narrow the margin was, Dowding gave permission to use 200 Polish and Czech pilots to bolster Fighter Command even though most spoke very little English.[94]

PHASE FOUR -- GOERING'S "GOOF"

With all the momentum going for the Luftwaffe, the RAF desperately needed a "turnover". "Der Dicke", the Fat One, Goering following a directive from Hitler, promptly fumbled, and gave it to them. He

switched targets from the sector airfields to raids on London. It was a blunder which changed the history of the world. Caught unawares, the RAF did poorly as the Germans "blitzed" London, but with a week's respite, Fighter Command rebounded. On September 15, a supposedly dead RAF rose from the grave and gave the Luftwaffe the telling blow of the battle and perhaps of the entire war. After the RAF 11th Group chopped the attacking Germans on the way in, the 12th Group "Big Wing" finally intercepted the bombers just as the Me-109s were forced to turn for home. The Germans lost 60 aircraft, but the major impact was psychological. The Luftwaffe crews thought that they had eliminated the Fighter Command and this clearly was not true. It was 15 September and winter with its rotten weather was coming. Air superiority still had not been achieved.

With the respite offered by the "Blitz" of London, the RAF Fighter Command was no longer threatened by the Luftwaffe and as long as air superiority was in doubt, no invasion could take place. Hitler and Goering had blown it. The invasion of Great Britain was going to have to wait. Although missions were continued against London, Operation Sealion was postponed.

SUMMARY

One of the great air battles of history was over. The Germans lost not by being overwhelmed, but by not being able to destroy the RAF Fighter Command. The Luftwaffe had to eliminate those swarms of

Hurricanes and Spitfires before an invasion of Britain, and it could not do it. There are a number of reasons this was true:

British Strengths:

• The overall command and leadership on the British side was better. It set an objective to defend Britain no matter the cost. The country, people, and leadership never faltered in their support of this objective. Air Chief Marshall Hugh Dowding had studied the air defense of Great Britain for years and had built a formidable defense system. The best possible man commanded RAF Fighter Command during the Battle.

• Because of the foresight of Air Chief Marshall Dowding and the unity of the British scientific community an effective air defense system was ready in 1940 to defend Britain. This system with its superb command, control, and communications capabilities was superior to the systems used by the Germans to attack Britain. Dowding's obdurate defense of the system from critics who did not understand the interception of aircraft was a key to its success.
• The technical capabilities of the Spitfire and Hurricane were suffi-

cient to meet the threat. During the Battle, the ratios of Hurricane to Spitfire was roughly 2-to-1 favoring the Hurricane. The Spitfire was the equal of the Me-109, while the Hurricane was largely inferior, but the Hurricane was deadly to all other German aircraft.

Using one day in the battle (Aug. 18),

the kill ratios of German aircraft by the Spitfire and Hurricane were very close to the same; i.e., about twice as many German aircraft were shot down by Hurricanes as by Spitfires. But the loss ratios were three times as high for Hurricanes; i.e., you were three times more likely to be shot down in a Hurricane as in a Spitfire.[95]

Right: The first of the U.S. follow-on fighters to the P-40 was the Lockheed P-38, big, fast and heavily armed, it was designed as an intercepter. It suffered from teething problems with its turbo-supercharged Allison engines. Very successful in the Pacific when refined, it suffered early in Europe with engine problems.

• British fighter production exceeded that of the Germans by a factor of nearly two-to-one throughout 1940.

• British intelligence was superb. The contribution of Ultra cannot be overemphasized. Dowding largely knew what the Germans intended throughout the battle. He also knew how well the RAF and the Luftwaffe were doing day by day.[96]

British Weaknesses:

•There was diversity among the leaders of the RAF about how to fight the battle. The "Big Wing" controversy with Leigh-Mallory did not help nor did Dowding's diffidence in handling the "Bomber Mafia" within the upper echelon of the RAF. His relief was ungrateful, rude, undeserved, and the worst kind of injustice. Thank God he was the man he was and was not relieved until the battle was over.

• On the other hand, the RAF remained on the pure defensive throughout the battle. There was no campaign to attack German airfields while they were mounting their raids or during the nights before. This could have disrupted maintenance, crew rest, and resupply, all of which would have helped RAF immensely. There were forces available for such a campaign yet no historians seem to fault the British for this oversight.

• Historians have hammered the Germans for the short rangeof the Me-109 and not using external drop tanks. Neither the Hurricane nor the Spitfire had any more range than the Me-109. Drop tanks would have been very useful for the Spitfire and Hurricane missions where time to climb was not critical. The form up of the 12th Group "Big Wing" is an example. With drop tanks, these wings could have been scrambled early enough to get into the fights over southeastern England.

• British fighter doctrine and tactics were faulty and caused many losses until they were abandoned.[97] The RAF proper did not develop the "Finger Four" variation of the Rotte and Schwarm until the spring of 1941.[98]

• The RAF, like every air force in the world, except the U.S., did not train enough fighter pilots. A shortage of trained, experienced pilots nearly cost the RAF the battle.

German Strengths:

• In 1940, the Luftwaffe was the largest, best trained, and experienced air force in the world. In particular, their fighter tactics were the best in the world.

• In the Me-109E fighter, the Germans had an aircraft that could dominate the skies performance-wise. The Germans had about 200 more Me-109s than the British had Hurricanes and Spitfires.[99]

German Weaknesses

• German leadership at the top suffered in comparison with the British. Goering was no match for Dowding, his fighter aircraft, his radar, command, control, communications, and the advantage of Ultra. Nor could he match Dowding's indomitable spirit.

• There was no coherent operational plan for the defeat of the RAF and the objectives kept shifting. Dowding and Park out-generaled their adversaries on the German side. Dowding never changed his objectives or methods while the Germans vacillated all over the place.

• Although the Luftwaffe Leaders supported Douhet and believed in first gaining "command of the air", they erred seriously in that they adopted some of the Principles of Douhet that were wrong and did

not adopt some that were right:

1) They, like Douhet, underestimated the importance of the fighter in gaining air superiority.

2) They, like Douhet, overestimated the effects of bombing on civilian populations.

3) They, like Douhet, felt that bombing airfields would gain air superiority.

4) They, unlike Douhet, reverted to Army Support; i.e, interdiction and close support (not Strategic Air) once they gained air superiority. It was up to the Wermacht to win strategically. They consequently never built a strategic air force with the range and payload to do the job.[100]

• Poor command decisions by the Luftwaffe top leadership:

1) The decision by Goering to close escort the bombers and Me-110s.

2) The decision to stop bombing the sector airfields and switch to London.

• German intelligence was unbelievably bad. There were any number of major failures by German intelligence:[101]

1) Ultra.

2) Poor knowledge of the British air defense system:

a) Radar and its vulnerability.

b) Location and vulnerability of the fighter operations centers.

c) Failure to initially locate and attack fighter bases (hit Bomber,

Coastal Command, and even Training Command bases).

3) Underestimation of British fighter strength and production throughout the campaign. Failure to locate and attack fighter production plants.[102]

• The poor performance of the Ju-87 and Me-110 aircraft.[103] They were "meat on the table" for the Hurricanes and Spitfires.

• The lack of range of the Me-109 and failure to design a viable external drop tank system for this aircraft.[104]

• Failure to develop and produce the Heinkel series of fighters (He-112, He-100) in conjunction with the Me-109 which could have been ready in 1940 and which had over a third more range.[105]

• Failure to produce a suitable strategic bomber with sufficient range, payload, and armament.

• Poor radio communications. Radios were not standardized between German fighters and bombers. Often escorting fighters could not talk directly to the bombers they wereescorting. Ground relay was often required to communicate.[106]

CONCLUSIONS

The Battle of Britain was one of the decisive battles of WW II and the first all-out air battle of the war. It also ended several myths of the time:

• The bomber does not always get though.

• The Luftwaffe was not invincible.

Unable to gain air superiority, Hitler's mind turned toward the East and the Soviet Union as he was forced to forget about the invasion of Great Britain.

THE PACIFIC

In the Pacific, the Japanese continued her war with China, and she became even more truculent and aggressive. In 1939 she introduced the Type 97 (Nate) into China and Manchuria. Claire Chennault dutifully forwarded the Nate's performance specifications to the U.S. War Department. Back came an engineering analysis saying his figures were impossible. Chennault knew better. These data were based on flight tests he had conducted on a captured Nate.[107]

In 1940, the Japanese introduced the Type 0, or "Zero" (Zeke). This new carrier fighter was the best in the world in 1940 and swept the skies in China. It had a 1,900 plus mile range with external drop tanks and could outperform every other fighter in Asia at this time.[108] When Claire Chennault forwarded the Zero's performance specifications to the U.S. War Department, they were again disregarded as it was felt he was exaggerating to get help for China and there was widespread disbelief that Japan could produce such a fighter.[109] This was criminal. U.S. pilots in the Philippines a year later had never heard of the Zero, until they died under the Zero's guns. From 1940 until 1942 this aircraft dominated the skies of China, Southeast Asia, and the Pacific (except for the AVG, and some individual U.S. Navy fights). *Note: Most of the fighters opposing the AVG were Hamps and Oscars not Zekes and were wrongly called "Zeros" by the AVG.

PART TWO--1941-1943

BALKANS INVASION

In 1941 Hitler's attention was directed toward Russia and his plan to invade the Soviet Union, "Operation Barbarossa". Before his armies were able to move (due to a late thaw in Russia) trouble arose in the Balkans and Greece. In November 1940, the Italians had invaded Greece only to be thrown back into Albanian by the tough Greek Army. The Italians had also invaded Egypt from Libya in September. The Italians had all they could handle in North Africa alone. Hitler was concerned about Il

Duce's over-extension into Greece and began to plan an operation to relieve the Italian Fascists. To accomplish this, the Germans had to obtain a clear route across Yugoslavia. The German Foreign Office negotiated clearance for the Wermacht on March 25, 1941. The pro-German Croats were pleased, but the pro-Allies Serbs were so upset that they launched a successful coup d'etat on March 27, 1941 and denied the Wermacht passage. Hitler was incensed and invaded Yugoslavia.[110]

The British stripped their forces in North Africa over General Wavell's objections and went to Greece's aid with inadequate forces. When Yugoslavia fell to the German

Blitzkrieg, the British/Greek positions were outflanked. Greece fell rapidly thereafter, and the British Army faced another Dunkirk.[111]

The Luftwaffe deployed into the Balkans and maintained air superiority throughout the campaign. Once again demonstrating their excellence in the support of Blitzkrieg, the Luftwaffe organization also exhibited its flexibility and outstanding ability to operate from forward deployments. The Luftwaffe was superb in this role and with air superiority, the Stukas destroyed position after position until Greece fell. For the third time, the Royal Navy was forced to disembark a British Army in the face of Luftwaffe air superiority

Right: The P-47 "Jug" followed the P-38 Lightning in the E.T.O. Built by Republic as a high altitude fighter, the rugged Thunderbolt with its huge radial engine and eight 50 caliber guns proved to be one of the premier ground attack fighters. Shown is a P-47N used late in the war mainly in the Pacific.

and this time losses were high.[112]

The British in North Africa did not recover for months after the embarkation to Greece, and in the interim, the German Africa Corps, commanded by Erwin Rommel, landed in Africa to stiffen the Italians in Cyrenaica. The British were to pay heavily for their Greek adventure and it was the air superiority provided by the Me-109 that solicited this exorbitant price.

It was then Crete's turn to face assault from the air dominated by the Luftwaffe. Warned by Ultra, the British built up the forces on Crete as much as possible, but there were just a few British fighters available for air defense. Rather than lose these, they

were withdrawn to Egypt. With total German air superiority, any movement in daylight by the British was nearly impossible. Under this umbrella, an airborne attack was staged in broad daylight by German paratroops. The German losses were horrendous, but with an extraordinary effort by the crack German paratroops, the airbase at Maleme was captured. From there, the whole island was taken. The Royal Navy lost two cruisers, four destroyers, and practically every other committed unit of the Mediterranean fleet was damaged by air attack. With the outstanding support of the Luftwaffe, Crete fell.[113]

It was a pyrrhic victory and the death of major German airborne operations for the rest of the war. Parachutists, gliders, and transport aircraft landed under direct fire from British Commonwealth troops. Losses were staggering. The Luftwaffe lost over 200 aircraft, mostly Ju-52 transports. Hitler was so shocked by the airborne troop losses that he never again attempted large-scale airborne operations. Few troops in the world could have pulled off this victory after taking the severe losses they encountered on the first day.[114]

GERMANS INVADE THE SOVIET UNION

On June 22, 1941 the Wermacht invaded Russia. On that day, the Luftwaffe destroyed 1,200 Russian aircraft. The Russian Air Force consisted of about 7,000 mostly obsolete aircraft. The tactical surprise achieved by the Germans is difficult to explain. The Russians had extensive warnings that the attack was

coming. The British had extensive intelligence from Ultra that the attack was coming, even as to when and where. The British did not disclose to the Russians the source of the intelligence (in fact they went to great lengths to conceal it) but they supplied accurate information to the Russians and there was sufficient corroboration to alert Stalin. Why he did not alert his forces remains an enigma. The tactical surprise achieved was not deserved but never-the-less complete.[115]

The Luftwaffe dutifully provided air superiority over the battlefields of Russia during 1941 and 1942. With about 2,500 first-line aircraft (1,500 remained in the West) this was a prodigious task.[116] The primary Russian fighter was our old friend from the Spanish Civil War, the I-16 Moska, and as we know, not the equal of the Me-109, or the new FW-190A. But, as we shall see, there were new and better Russian fighters coming, and lots of them.

In 1942, Goering promised Hitler he could resupply the Stalingrad pocket by air. In the face of Russian air superiority (regained by swarms of Russian fighters) and rotten Russian winter weather, this proved impossible. Needing 600 tons of supplies per day to supply the surrounded German Sixth army, the Luftwaffe actually averaged less than 100 tons per day. The pocket collapsed as did German air superiority for the rest of the war.[117]

AIR WAR IN THE WEST -- 1941

During the winter of 1940 and the

spring of 1941 the British tried to rebuild its Fighter Command with improved aircraft. Spitfires replaced the Hurricanes and the night bomber campaign began in earnest against Germany and the occupied countries.

Night Fighting

Britain installed radar in Blenheims and Beaufighters to intercept German night bombers. This was the beginning of the all-weather/night fighter. These night fighters steadily improved and as they did, viable all-weather fighters became a reality. In 1940 neither the Luftwaffe nor the RAF was ready for this type of warfare. Crews were not trained, there was no recovery equipment for night fighters to return and land, and there were initially no suitable airframes to perform night fighter duties. With the predominantly poor flying weather in Europe, crew instrument proficiency requirements were very high.[118]

The night-fighter had entirely different requirements than a day-fighter. There was less requirement for maneuverability as the night-fighter was designed primarily to oppose bombers. It had to be stable for instrument flight, fast enough to overtake the bomber, be able to handle the radar payload, have enough fuel for extended loiter, and be heavily armed as it would probably have only one shot. There would be no second passes. In most cases, this meant a fast, twin-engine fighter or an adapted light bomber (e.g., Me-110, He-219, A-20 [P-70], P-61, Blenheim, Beaufighter, or Mosquito).

The idea was for a ground control radar to vector the fighter into a range where the night-fighter could take over its own intercept using its airborne radar. The radar operator then would talk his pilot in close enough for the pilot to acquire the target visually, aim, and fire. With their rudimentary radar sets, all of this was really difficult. Overtake, turns, and horrible weather could really make this a most difficult and dangerous task. Possible mid-air collision and spatial disorientation were a way of life.[119] But slowly the equipment got better and the night-fighter became a viable weapon.

Enter the "Wild Boar"

By 1943, the Germans were shooting down 10-20 % of the British night bombers routinely. "Wild Boar" missions were being flown by some pilots in day-fighters with special landing equipment to successfully recover and land at night. These fighters received no radar vectors but sneaked into the big British bomber streams using visual contact only. To intercept blacked out bombers with one's eyeballs was "hairy" at best. These fighters had mixed results. The extensive bomber streams made for plentiful targets. Of the nine awards of the Swords and Diamonds/Oak Leaves to the Knight's Cross, (the Nazi's Victoria Cross) two were to night-fighter pilots.[120]

Day Fighting

In the spring of 1941, the Spitfires began day attacks on ground targets in France and the low countries in what they called "Rhubarbs". During the summer months, a new, really

fast radial-engined fighter opposed them. It was swift and maneuverable. At altitudes around 20,000 feet, it was superior to the Spitfire V, the latest variant being used by the RAF.[121]

The RAF was really curious about this fighter but details were lacking. It was the only new single-engine prop fighter to achieve mass production in Germany after the war started. But unknown to the British, the new fighter was having the typical teething problems of any new fighter. The BMW radial, fitted with a close fitting cowl had serious cooling problems which put severe stress on the lower cylinders of the second row of this twin row engine. The oil system gave trouble, as did the new constant speed prop. Engine and propeller problems caused more losses than operations. The airframe, however, was outstanding.[122]

The FW-190A was refined and by early 1942, it was the best fighter in the Luftwaffe. It was a dream to fly, easy to land and structurally sound (the Me-109F was having serious structural problems with both tail and wings). The wide landing gear made the ground loops of the narrow-geared 109 a thing of the past when units were re-equipped with the FW-190.[123]

U.S. ENTRY IN THE WAR

On Sunday, December 7, 1941 the Japanese Navy bombed Pearl Harbor. Hitler and Mussolini declared war on the U.S. two days later. In his memoirs, Winston Churchill says that he breathed a sigh of relief as he no longer feared the war's outcome.[124] Americans were dying and we were again in an "all-out, no holds-barred, gut-busting" world-wide war. Today, most Americans have forgotten how close those dark days were as the Allies lost battle after battle.

The LINEUP -- ALLIES Vs. AXIS DECEMBER 1941

It would be useful to reconsider the fighters of the world again as of December 7, 1941: (note: all engines are liquid-cooled unless listed as a radial).

• RAF – Spitfire V, Rolls Royce Merlin 45 of 1,470 HP; 374 MPH, Armament varied, 8 x .303 wing-guns, or 2 x 20 MM cannon and 4 x .303 wing-guns, or 4 x 20 MM cannons; Development of the Spitfire I that fought the Battle of Britain.[125]
• Royal Navy – Sea Hurricanes, Rolls Royce Merlin II of 1,030 HP, 316 MPH, 8 x .303 caliber wing-guns; Carrier development of Battle of Britain Hurricane I. Obsolete but better than Skua and Gladiator.[126]

o Luftwaffe – FW-190A, BMW 801D radial of 1,700 HP, 401 MPH, 2 x 20 MM cannons, and 2 x 7.9 MM machine guns all synchronized. 2 x 20 MM cannons in the wings. Arguably the best fighter in the world at this time. Fast, maneuverable, (one of the fastest rolling fighters in the world).

• Soviet Union – Yak-1, Klimov M-105PF-1 of 1,260 HP, 364 MPH, 1 x 20 MM, 1 x 12.7 MM. Very maneuverable and fast at low altitude. Reliable and developed into a series of improved fighters later on (Yak-[3-9]).[128]

• Regia Aeronautica – Macchi 202 Folgore, Daimler Benz 601A of 1,175 HP manufactured under license by Alfa-Romero, 370 MPH, 2 x 12.7 MM, maneuverable, fast and a delight to fly. Superior to Tomahawk, Kittyhawk, and Hurricanes it faced. Equal to Spitfire Vs. With this aircraft the Italians rejoined top fighters of other nations, but it was dependent on DB engines which were in short supply. Just a few produced.[129]

• Japanese Army – Nakajima Type 1, Mod 1 (Ki-43 – Oscar), Nakajima Ha-25 radial of 990 HP, 306 MPH, 2 x 12.7 MM machine guns. Very maneuverable but under-powered and lightly armed. Effective in 1941-42 when Japanese had overwhelming air superiority. Often confused with the Zeke.[129]

• Japanese Navy – Mitsubishi Type 1, Zero (Zeke), Nakajima Sakae Radial of 950 HP. 331 MPH. 2 x 20 MM cannon and 2 x 7.62 machine guns. Best carrier qualified fighter in the world in 1940-42. Extremely maneuverable and longest ranged fighter in the world at this time.[131]

• USAAF – P-38E, 2 x Allisons 1,150 HP each, 395 MPH, 1 x 20MM cannon, and 4 x .50 caliber machine guns concentrated in the nose. Just coming into service. Many teething problems especially with engines at high altitude in Northern Europe.

• U.S. Navy – F4F-3, Carrier qualified, Wright Cyclone, of 1,200 HP, 330 MPH, 4 x .50 caliber machine guns, maneuverable, rugged, and well armored.[133]

Not surprisingly, fighters were improving. The U.S. and USSR were still playing catch-up. Superiority had switched from the Spitfire V to the Focke Wulf 190, but the Spitfire IX was just around the corner. The Me-109F was on-line and significantly outperformed the Me-109E, but ran into structural problems with both the wings and tail and was inferior to the FW-190.

CONFLICT IN THE LUFTWAFFE

At the top of the Luftwaffe, two of Goering's key deputies, Udet and Milch were in huge arguments about the viability of the Me-109 versus the FW-190. Udet, the old fighter ace from WW I, favored the FW-190. Milch favored the continued concentration and expansion of production on the Me-109. Udet was being blamed heavily for Luftwaffe failures during the Battle Of Britain (Milch was suspected) and repeatedly took his problems to his old friend from WW I, Goering. Udet was not up to his high administrative post and was on the verge of a nervous breakdown (accelerated by amphetamine and alcohol abuse). Milch won the day by putting forward what he called "The Goering Plan" as expansion for the invasion of Russia (expanding the Luftwaffe by a factor of four). In that plan he ordered the all-out production of Me-109s (top priority) to the detriment of the FW-190 and the budding Me-262. That decision might have cost Germany the war.[134]

Udet thought so, as he killed himself leaving a grotesque message on the wall that the "Iron Man" had failed him by selling out to Milch. Iron Man was Udet's nick name for Goering. Nearly everyone else called him the "Fat One". Werner Molders was killed flying to Udet's funeral to be one of his pall bearers. [135]

This decision to go for quantity rather than quality has been made before. The British made it in 1917 – the result was "Bloody April". The Germans made it repeatedly during Hitler's reign by stopping R & D, to push production. Failure to develop the He-112 line is an example. But the real failure was slowing the development of the Me-262 which Milch, Goering, and Hitler failed to comprehend until too late. The Me-262 was to become operational in 1944 when it was too late, very much like the Fokker D VII which arrived too late in 1918. USAAF bombing of the Messerschmitt factory in 1943 helped in this delay, as it destroyed a number of critical Me-262 jigs needed for its production. Hitler further delayed the fighter by four to six months in insisting it be made into a bomber. These delays coupled with the technical problems in developing the jet delayed operational service until it was too late. [136]

Interestingly, Air Marshall Milch was to be relieved for protesting Hitler's decision to make the Me-262 a bomber. He is reputed to have said "Any small child can see this is a fighter not a bomber". He was immediately dismissed as Luftwaffe chief of aircraft procurement. [137] Perhaps Ernst Udet had the last laugh after all.

If the Germans had concentrated on the Me-262 and got it operational by 1943 our history might have been very different. Certainly the invasion of Normandy would have been in doubt and the Strategic Bombing Offensive battle over Germany would have been lost until the British Meteor and Lockheed P-80 could have been deployed. And those aircraft had straight wings! The swept-wing technology that delayed compressibility was incorporated only in the Me-262. This allowed the Me-262 to approach the speed of sound while the Allied straight-winged aircraft were slower.

It's all speculation, but with air superiority on both fronts, the Germans might have prevailed. Could we have delivered an Atomic bomb with B-29s against an all-jet fighter air force? It is scary speculation. Perhaps a Divine Providence saw to it that Hitler, Goering, and Milch did not realize what they had.

FIGHTER DEVELOPMENT AND PRODUCTION

Fighter development during the war was different in each individual country. Each belligerent was a little different in how they addressed the development and production of fighters.

While the Germans were to fight WW II with substantially only two prop-driven single-engine fighters (The Ta-152 was a development of the FW-190) and produced no really viable prop-driven twin engine fighters (the Me-110 was a failure as an air-to-air fighter and the Do-335 never really became operational), the Americans had many. We began with obsolete P-40s, P-39s, and F4Fs, but the U.S. had a string of new fighters coming. However, it would be 1943 before they would really impact (P-47s, improved twin engine P-38s, P-51s, F6F Hellcats, and F4U Corsairs). Unlike Germany, American development continued throughout the war, the twin engine P-61 night-fighter became operational in 1944, all operational fighters were improved, and the F7F (twin engine), F8F, P-82 (two P51s joined into one aircraft), P-80 Jet Fighter just barely missed the war. There were at least a dozen of other prototypes in development when the war ended. [138]

The American production plan had begun in November of 1938 [139] and the juggernaut that would build an average of 65,000 planes a year during the war became a reality early in WW II. In its biggest year, 1944, the U.S. would produce 96,318 aircraft of all types. [140]

The British went primarily to continually-improved Spitfires, with Whirlwinds, Typhoons, and Tempests following. They also used some American models; Mustangs, Martlets, Corsairs, Kittyhawks, and Warhawks were among the fighters provided by lend-lease. The British out-produced the Germans in fighters from the summer of 1940 through 1942. [141] After that, the Germans made the decision to produce more defensive fighters, particularly after the conflagration of Hamburg. [142] But even then, Milch increased fighter production to just 360 per month. [143] But a year later Germany was producing 2,300 single-engine fighters a month. [144] What would have been the result had German production peaked earlier?

British aircraft production of all types averaged 18,793 per year over the seven calendar years from 1939-1945.[145] It is interesting to realize that the democracies whose existence was challenged early-on made the hard economic decisions to disrupt their civilian economies very early in the war while both Germany and Japan did not go over to a war economy until much later.[146] In both cases, the Axis made their decisions too late to make a difference. Italian production was so low (2,000 aircraft per year) it never became a factor.[147] By building prototypes and putting them directly into production, the Russians were able to expedite production at the cost of operational efficiency. Many of the bugs of the prototypes were sorted by operational pilots in combat. Often this policy did not work out well for the operational squadrons and losses were high. It is doubtful that this practice would have been acceptable in western squadrons. But it did provide operational testing without hindering production. And since the Russian Army Air Force (VVS) was willing to accept the tremendous waste as an exigency of all-out war, it worked. Ineffective fighters were withdrawn from combat and modified, or discarded.

The VVS began WW II at a severe disadvantage. Stalin had purged the VVS of 70% of its top leaders in the late 1930s. In the early stages of the German attack, VVS aircraft losses had been enormous (the Russians themselves admitted los-

ing over 5,300 aircraft). The VVS Commander in Chief, General Smushkevich who was directly responsible for all dispositions was summarily executed.[148] But most of the aircraft destroyed were obsolete and replacement aircraft were more modern even though they were forced to be tested and developed in combat. By mid-to-late 1942, not only quantitative equality had been achieved with the Luftwaffe, but improved Yak and Lagg fighters were of equal or better quality than the Luftwaffe Me-109s and FW-190s they had to fight. Also, as most of the Russian aircraft destroyed in 1941 had been on the ground, there remained a cadre of experienced pilots to train new pilots.[149]

When one realizes the feat of Russian industry in this accomplishment, it is almost unbelievable. More than 100 aircraft factories and their work force had to be dismantled and evacuated to the east when Hitler invaded.[150] To achieve equality in the air by 1942 was an incredible feat for Russian industry. Even under these circumstances, the Russians averaged 22,603 aircraft a year from 1939-1945.[151] The quality of the fighter force was superior after Stalingrad in late 1942 and early 1943. The Yakolev and Lavochkin fighters could hold their own with the best of the Luftwaffe particularly at low altitude.[152] When one considers their superior numbers plus lend-lease fighter help from the U.S. (P-39s, P-40s, P-63s) overall air superiority passed to the Russians from the winter of 1942-43 on. In addition, the Germans transferred many experienced units back to Germany to fight the Allied bomber offensive.

Of all the belligerents in WW II, German aircraft production is perhaps the most incredible. Not because of its extraordinary effort, but because of the lack of it. As we have seen, during the height of the Battle of Britain only 140 Me-109s a month were being produced while the British were producing 500 Hurricanes and Spitfires.[153] Hitler thought the war in the West was over when France was defeated and he was unwilling to subject his civilian population to privation until the war turned against the Axis in 1943. The following table is mind-boggling (Germany was the aggressor in WW II and Hitler declared war on three-fourths of the world without maximizing Germany's production for war!):

**German Aircraft Production
1939-1945**
All-Types[154]

- 1939 —————— 8,295

- 1940 —————— 10,247

- 1941 —————— 11,776

- 1942 —————— 15,409

- 1943 —————— 24,807
- 1944 —————— 39,807

- 1945 —————— 7,540

Average yearly production = 13,652 aircraft.

Germany produced more aircraft by a factor of about a 50% increase during the heavy allied strategic attacks of 1943-44.[155] Critics of the Allied Strategic Air Offensive continually raise these German production increases to condemn the offensive as a failure. These critics take no account of a number of factors: 1) The German aircraft industry was very inefficient and disorganized until centralized production scheduling and control was instituted by Albert Speer in 1942. 2) The German war economy was idling until mid-1942. 3) Slave labor was introduced in large numbers. 4) Many large factories were moved underground. 5) Desperation is a prime motivator to both governments and their people.[156] Fortunately, due to the Allied raids there was no fuel for the additional planes the Germans produced and no experienced pilots to fly them. Both the lack of fuel and pilots were directly due to the Allied Strategic Air Offensive. It is also interesting to note that the British out-produced the Germans in every year except 1939 and 1944, a fact not generally known.[157]

Japanese economic planning was equally confused. Like the Germans, the Japanese maintained production at near-peacetime levels during the first two years of the war as they anticipated a negotiated settlement early in the war. Only after disasters began to culminate at the end of 1942 did Japan re-address their production capacity and capabilities. Given the state of Japanese industry, this was too late.[158]

When one considers the fact that the Japanese and Germans were the aggressors, these mistakes are difficult to understand. One of the first laws of operational planning is to prepare for all eventualities. It is obvious that the Axis did not adhere to this law and suffered the consequences at least in the air.

This is of interest as many contemporary observers at the beginning of the war attributed much of the success of the Axis powers from 1939-1942 to their dictatorial forms of governments. As hindsight verifies, the strategic planning of an oligarchy is only as good as the dictator or ruler. In WW II, the Axis were very poor planners strategically when compared with the Allies. The British made some serious tactical blunders to be sure (Norway and Greece immediately come to mind) but these mistakes were largely in tactical execution or an overestimation of British capabilities in the face of Luftwaffe air superiority.

Finally, as indicated earlier, Italian planning, production, and execution proved to be a detriment to the Axis throughout the war.

THE WAR IN THE PACIFIC

After the Japanese attack at Pearl Harbor, as Yamamoto predicted,[159] the Japanese ran wild all over the Pacific and Southeast Asia for six months to a year. His predictions turned out to be very accurate, as the Battle Of Midway occurred in June of 1942. The U.S. was never in serious trouble after that, as most of Japan's experienced carrier aviators were lost at Midway. Of course Yamamoto had no way of knowing that the Allies had broken the Japanese codes and knew his most secret plans ahead of time.

But for six months the Japanese were successful beyond their fondest dreams and it was the predominance of the Zero that was the main reason for these successes.[160] Although warned by Chennault and by evi-

dence from Russian experiences, the Allies were caught totally by surprise. The Zero's performance and range were major shocks. None of Chennault's reports had been forwarded to the field. In light of the detail of these reports which included photographs, complete details of encounters, and accurate estimates of performance and impact, this was criminal. British envoys also forwarded detailed reports to Australia and Britain with much the same result.[161]

The Russians, who were assisting Chiang Kai-shek in China at this time, fought the Zero unsuccessfully with I-16s and I-153s. They never had a chance. The Zeros outnumbered the Russian/Chinese fighters and could outperform them in every aspect of performance. They were faster, more maneuverable, climbed and dove better, and were flown by more experienced aircrews. The Japanese flew 529 sorties, some as long as 1,200 miles round-trip, destroyed 99 Russian/Chinese fighters with no air-to-air losses (two Zeroes were lost to ground fire). This was an incredible performance and observed by many Allied observers. Yet the Allies were surprised by the Zero's performance six months-to-a-year later.[162]

In 1927, Billy Mitchell predicted that Japan might attack the U.S. bases in Hawaii using carriers from north of the islands. On March 31, 1941, the joint air commanders, Major General Frederic Martin and Rear Admiral Patrick Bellinger in drawing up a joint defense plan, predicted as most likely to occur in Hawaii, a carrier attack launched from 233 miles to the north. They also predicted a joint attack with submarines as well as aircraft.[163] Evidently everyone forgot this estimate when they searched for the carrier forces after the strike on Pearl Harbor. The search went in the other direction.

Japanese Air Power(combined Naval and Army Air Forces) overwhelmed Allied forces in the Philippines, Guam, Wake, Malaya, the Dutch East Indies, and the Indian Ocean with relative ease. The inexperience of the Allied pilots, inferior aircraft, and overwhelming Japanese numbers left no doubt as to the outcome for the Allies. Even when in near equal strength, the Allied pilots were not tactically trained to deal with the swift, maneuverable, superbly-handled Zeros. They tried to dogfight with them and died. Only the AVG and the U.S. carrier fighters developed tactics during the first six months of the war to successfully fight the Zeros.[164]

It is interesting to note that both the P-40s of the AVG, and the F4Fs of the Navy used the pairs and the tactical spread Oswald Boelke developed in WW I in their own adaptations. Both the Navy tactics and Chennault's were independent of the Rotte and Schwarm, used by the Germans, but very similar. At the Coral Sea and Midway, the Fleet F4Fs held their own and using the Thach Weave and dive and zoom were successful.[165] After intensive training by Chennault, the AVG was successful from day one.[166]

The only worthwhile resistance

during 1941 and early 1942 was in the Philippines where the U.S. retreated into the Bataan peninsula. There they held out for five months, but our air forces were overwhelmed by superior numbers. At Clark Field most of our air forces were wiped out on the ground.[167]

The air commander was General Louis Brereton whose career during WW II was at best checkered. He escaped from the Philippines to command the U.S. Air Forces in the Dutch East Indies (debacle two). He then commanded 10th Air Force in the CBI where one of his units was the famous AVG under Chennault. Brereton respected Chennault as a solid combat leader. Reassigned to North Africa, he commanded the first raid on the Polesti oil fields from North Africa by low-level B-24s (another debacle). He then showed up in command of the airlift for the airdrops associated with Operation Market-Garden in Holland (debacle four). He was either incompetent or the unluckiest air general of the war.[168]

In August of 1942, the U.S. invaded two islands in the Solomons named Tulagi and Guadalcanal. One of the most valiant fights in the history of U.S. forces was about to begin as a handful of Marine and USAAF fighters fought the Japanese Navy to a standstill. They were nicknamed the "Cactus Air Force". The name was derived from the code name for Guadalcanal which was "Cactus".[169]

Operating from the airfield on Guadalcanal, "the Unsinkable Aircraft Carrier", the Cactus Air Force destroyed 96 Zeroes, 92 Bettys and 75 other types, mostly float planes

while losing 101 friendly planes and only 38 American pilots killed. Japanese records indicate that between August and November 1942, they lost 344 fighters, 125 medium bombers, and 198 torpedo and dive bombers in the Solomons. Most of these had to have fallen to these valiant Americans flying F4Fs, P-400/P-39s, and P-38s flying from Guadalcanal as there was no other combat action in the Solomons at this time.[170] The fighting on Guadalcanal was particularly bitter. The air base there was under constant attack by land, air, and sea.[171]

On April 18, 1943, this group of pilots culminated their actions by intercepting Admiral Yamamoto, and shooting him down in flames. This flight of P-38s included several veterans of the 67th Fighter Squadron who had upgraded from the P-400 to P-38s. Major John Mitchell from the 67th led the flight.[172] There were two Betty bombers and six escorting Zeros. Both of the bombers were shot down. There is still debate about by whom. Captain Thomas Lanphier and Lt. Rex Barber both claimed credit for Yamamoto's aircraft. Time has given weight to the claim by Barber. Whatever else, the dead reckoning navigation by Mitchell for almost three hours over open water was outstanding and the mission accomplished.[173]

Much of the credit for the success of the Cactus Air Force must go to their tactics which were based on Chennault/U.S. Navy doctrine of fighting pairs and dive and zoom tactics. The Americans attacked from advantage and dove out of the fight always maintaining the initiative and refusing to dogfight with the Zeros. The Japanese were also at disadvantage as they were a long way from home. On some occasions the Japanese fighters were bounced when they still had their external drop tanks and to jettison them meant they could not return to their bases at Rabaul or Kavieng. These fighters were easy meat for the F4Fs and P-38s (P-400s and P-39s were used as close-support fighters since they performed poorly at altitude. The P-400 was a lend-lease version of the P-39 rejected by the British and commandeered for U.S. use. Like the P-39 without superchargers, the P-400 was worthless above 15,000 feet).[174]

In late 1942, the war in the Pacific split into two distinct commands with General MacArthur in charge of the Southwest Pacific and Admiral Nimitz in charge of the Central Pacific. By 1943, the fighter war split also as the P-38 Lightning became the primary USAAF fighter in the Southwest and the F6F Hellcat primary off the carriers in the Central Pacific. The P-38 was supplemented by the P-47 and the F6F by the F4U. The F4U was used as a land-based fighter only. Interestingly, the British Navy used the F4U aboard ship nine months before the U.S. Navy used it on carriers. Initally, the F4U was considered unsuitable for carrier operations by the U.S. Navy until April 1944 when it began to replace some F6Fs. The first operational F4U Marine squadron arrived in Guadalcanal in February 1943.[175]

All of the fighters mentioned above were superior to anything the Japanese could put into the air for

the rest of the war. From 1943 on, the U.S. had overwhelming numerical superiority in fighters as well and none of the improved Japanese fighters could offset this superiority. In effect, the air superiority war ended in 1943 in the Pacific. The quality of Japanese pilots and aircraft was relatively gone and led to the drastic influx of the Kamikazes in 1944. But the search for air superiority was rough in 1942 and the outcome much closer than many realize. The Japanese were headed for Port Moresby, on the Southern tip of New Guinea as a prelude to invading Australia. General MacArthur was determined to stop them. General Marshall, informed that MacArthur had relieved the American Air Commander General George H. Brett, assigned General George C. Kenney to the job to replace Brett. MacArthur had a reputation of not getting along with his air deputies. This time was different.

Kenney had two kills as a fighter pilot in WW I and extensive experience particularly in the development of aircraft and maintaining them. He was a practical air commander who knew his stuff. He was perfect for a jungle war where his air force had no supplies, was disorganized, and most of all needed a competent, confident commander. MacArthur trusted him from their first meeting and Kenney never let his commander down.[176]

Kenney knew the USAAF inside out. He brought with him two outstanding subordinates, Brigadier General Ennis Whitehead to command his fighters and Brigadier General Kenneth Walker to command his bombers. He relentlessly sought to improve his Air Force so it could

take the offensive. He replaced tired and depressed officers with men who wanted to fight and get the job done. Younger men were promoted on merit and he rewarded aggressiveness and innovation.[177]

Some of his innovations were mind-boggling. A-20 light bombers and B-25s were equipped with between eight to fourteen .50 caliber machine guns and became "strafers". A new bomb delivery was developed for hitting ships. By flying low at 50-100 feet his bombers would drop bombs with delayed fuses onto the water. The bombs then would skip into the side of the ship and detonate. Using the forward firing fire power of his strafers to keep enemy heads down, this type delivery was extremely accurate and effective. He promoted the installation of a 75 MM cannon in a B-25, which was put into production.[178]

Kenney also used his four-engined B-17s as skip bombers against shipping at night.[179] His aircraft usually attacked under flares dropped by another aircraft. Kenney knew his way around the USAAF and was close to General Hap Arnold, the Chief of Staff of the Army Air Force. When he discovered that the ETO Commanders preferred the P-47 and P-51 fighter aircraft rather than the P-38, he contacted Gen. Arnold to send him all the P-38s produced. He pointed out with two-engines and its long range it was ideal for the over-water island campaigns with which MacArthur's forces were faced. He also requested the

B-24 for his theater as it had a longer range than the B-17. In the main his requests were granted.[180]

Weldon Rhoades in his book, *Flying MacArthur To Victory*, gave credit to General Kenney as follows: "By accomplishment and insight, he helped convince MacArthur that command of the air was an absolute essential for successful, sustained ground and seaborne offensives." MacArthur himself had this to say in his autobiography:

"Major General George C. Kenney came to command the Allied Air Forces. Of all the great air commanders of the war, none surpassed him in those three great essentials of combat leadership: aggressive vision, mastery of air tactics and strategy, and the ability

to extract the maximum in fighting qualities from both men and equipment. Through his extraordinary capacity to improvise and improve, he took a substandard force and welded into a weapon so deadly as to take command of the air whenever it engaged the enemy, even against apparent odds."[181]

Kenney's Fifth and Thirteenth Air Forces gained air superiority over the islands of the Southwest Pacific in 1942 and never lost it. The P-39s, P-40s, P-38s and P-47s gained air superiority and his bombers maintained it from 1942 until 1945. During that time, the P-38 shot down more Japanese aircraft than any other fighter in the Pacific.[182]
The "Jungle Air Force" using its air superiority gained literally on a "shoestring" by P-40s, P-39/400s, and P-38s was decisive in the Papua

Top: *The Me-262 was the first jet fighter to be operational. It's performance exceeded anything flying. Fortunately, it arrived too late to turn around the war.*

Campaign in late 1942. General Kenney's air forces virtually isolated the Japanese advancing on Port Moresby by destroying their tortuous supply lines across the Owen Stanley Mountains. Then his air force flew large U.S. Army and Australian units into the Gona and Buna area. Buna fell to the Allies on January 2, 1943, and Sanananda on January 16.[183] With very few resources, Fifth Air Force had won a major victory over a tenacious enemy.

Using Kenney's innovative, land-based airpower, MacArthur advanced up the coast of New

Guinea by-passing Japanese bases by amphibious landings near the Japanese without facing their opposition. As objectives, he chose harbors and airfield sites in relatively undefended areas that effectively neutralized the by-passed Japanese facilities (e.g., Lae and Finschhafen). These sites required a high level of organization, logistics, engineering, and construction capabilities to build air bases and facilities rapidly in the middle of nowhere. This was possible only under the umbrella of air superiority provided by Kenney's P-38s. The overall objective was the isolation of the huge Japanese base at Rabaul located on the eastern end of the island of New Britain. Admiral Halsey was closing on Rabaul by advancing along the Guadalcanal, New Georgia, Bougainville, New Britain route. Halsey used the same bypass technique as MacArthur and the two forces formed a pincers movement isolating Rabaul.[184]

It was along these routes that most of the major air battles of the Pacific War occurred as the Japanese reinforced Rabaul on New Britain and Kavieng on the island of New Ireland repeatedly. These two bases were the point of repeated carrier attacks by Halsey and long range heavy bombers from Kenney's forces. Navy F6Fs, Marine Corsairs, and U.S. Army P-38s and P-47s were met by Japanese Tonies, Oscars, and the venerable Zero. The U.S. forces by 1943 had qualitative and numerical superiority in fighters and were slowly grinding the smaller inferior Japanese forces down. Most of the Pacific USAAF and U.S. Marine Aces

flew in these campaigns. In December 1943, the U.S. invaded Eastern New Britain, the other end of the island from the huge Japanese base at Rabaul.[185]

All of these amphibious operations required air superiority to take place. If the air commanders could not provide air superiority, the attacks did not go.[186] When asked to provide continuous air cover over the beaches, Kenney refused (he did not have enough assets to do so), but insisted instead on using his aircraft to attack and reduce the Japanese air capabilities thereby providing air superiority.[187] All of the landings were successful.

THE MEDITERRANEAN

When the Italians entered the war in 1940, they sent an army under Graziani to Libya to attack the British in Egypt under Wavell. After the Italians crossed the frontier into Egypt, Wavell decisively counterattacked. The Italians retreated and the British pursued aggressively using their air superiority and naval power in conjunction with well-handled British armor. During the pursuit, the Italian Army largely disintegrated and over 130,000 Italians surrendered. Much of this disintegration was caused by British air attacks using American Kittyhawks, Warhawks, and Hurricanes along with light bombers during the retreat. The British had air supremacy all during Wavell's counter offensive. They were about to apply the "coup de grace" to Graziani when Churchill ordered a large part of these forces to Greece to protect the Greeks from the invasion by the Nazis.[188]

With the largest part of his forces deployed to Greece on what proved to be a "wild goose chase", Wavell was faced with Italian reinforcement by the Germans under an unknown German commander named Rommel. Warned by Ultra, Wavell was aware that he was outgunned and outnumbered. Rommel was well-equipped and supplied. The "Brits" in Cyrenaica were about to learn the lessons of "Blitzkrieg" the hard way. Rommel broke through the thin British lines and pursued them back to the Egyptian border. British intelligence knew little about Rommel, but it would not be long before they knew him well.[189]

But Rommel was unable to take Tobruk. Malta-based aircraft and the British Navy using Ultra, repeatedly clobbered and interdicted shipping and supplies bound for the Africa Corps. Without these supplies, Rommel was actually overextended, with a poor supply situation on the Egyptian Border, and large British forces hundreds of miles in his rear at Tobruk. General Paulus was sent from Berlin to appraise Rommel's situation. Paulus found Rommel's condition perilous and broadcast his findings to Berlin by radio. Paulus's report was read by Churchill through Ultra and he goaded Wavell to attack. Wavell, over his own better judgement, did attack and was bloodily repulsed by the capable Rommel. Wavell was replaced by Auchinleck. Auchinleck refused to attack until he rebuilt his forces and regained air superiority. When he did attack, Rommel had to fall back and shorten his supply lines as the British again went over to the offensive.[190]

Realizing Malta was the problem, both Kesselring (who now commanded in Italy) and the Italians wanted to reduce Malta to ashes by bombing and then to take it. Hitler wanted the Italians to take Malta. He reinforced the "Med" with II Corps, five Gruppen of Ju-88s, one of Ju-87 Stukas; one of Me-110s and the experienced JG 53 with four Gruppen of Me-109Fs. These aircraft were to reduce Malta which was a real thorn in Rommel's supply side.[191] JG-27 was already deployed in North Africa itself.[192] General Student, who had commanded the invasion of Crete, was set to planning an airborne invasion of Malta and collecting the necessary resources.

During the period March 20, 1942 until April 28, 1942, German bomber and fighter Gruppen flew more than 11,500 sorties against Malta and the bombers dropped approximately 7,213 tons of bombs on that very small Island. [193] Malta was a Guadalcanal of the Mediterranean as these Gruppen took grievous losses attacking this unsinkable "aircraft carrier". The British kept reinforcing Malta with fighters flying in Spitfires from carriers on April 20th and May 9th. The April 20th aircraft were knocked out as soon as they landed, but the 67 Spitfires flown in on the 9th were placed in cave shelters and refueled and rearmed in minutes. The British were aided again by the German high command insisting on the close support of German bombers by their fighters. As in the Battle of

Britain, this policy hamstrung their fighters. Slowing their fighters down to closely escort their bombers put the Luftwaffe fighters at a severe speed disadvatage versus the British Spitfires. It was a disadvantage they could ill afford. German losses skyrocketed on May 10-12 as British Flak and Fighters resisted strongly. The Luftwaffe II Corps had just reported that Malta's defenses were all but wiped out. [194] It was like a reprise of September 15, 1940 during the Battle of Britain.

Student and his XI Corps, reinforced by two crack airborne Italian and 70,000 seaborne Italian troops, had collected 500 Ju-52 transports and a flotilla of gliders. In June 1942, just as Rommel launched his second offensive, Student was summoned to Rastenburg where it became apparent that Hitler was afraid of the British Navy's control of the Mediterranean and never intended to invade Malta. Meanwhile the British built up the fighter defenses of Malta.[195] It was never taken. With Ultra and Malta in British hands, Rommel's supply lines were doomed. Seaborne or airborne operations required air superiority and the Germans never gained it and held it over Malta.

As Rommel's ground forces overran the forward British airbases, during his second offensive, the Luftwaffe gained temporary air superiority over the battlefields in North Africa with his Italian Fighters and two German fighter Geshwaders. Using this superiority, Rommel took Tobruk in 28 hours using Stukas and other close support aircraft to blast through the fortress defenses. Flying with JG 27, a fighter Geschwad-

er that helped gained that air superiority, was Hans-Joachim Marseille. Marseille was thought by many to have been the greatest ace of them all. He ran up 158 victories in only nine months. He claimed 17 victories in one day. A check of British losses that day indicate that he probably did not exaggerate. Most of his victories were fighters. To a man, his Gruppen mates say that he was the most extraordinary gunner they ever saw, particularly in his ability to deflection shoot.[196] (accurately fire on turning, maneurvering air-to-air targets).

Rommel advanced on to El Alamein where he was finally defeated by General Bernard Montgomery in November. Montgomery with Ultra operating well, knew Rommel's strength and dispositions. The British airpower at Malta, reinforced and as active as ever, interdicted Rommel's supply lines across "The Med". While Rommel was receiving little or no resources or reinforcements, Montgomery was receiving heavy reinforcements. The British were very near their supply bases while Rommel's bases of supply in Libya were hundreds of miles away. Because of Malta, Rommel's Africa Corps was receiving a mere trickle of what it needed. Time was on Montgomery's side. Characteristically, as Rommel closed with the British in August and knowing that his time was short, he attacked.[197]

Unfortunately for Rommel, Montgomery knew his plan through Ultra and mousetrapped him with intricate defenses. Montgomery's air had been heavily reinforced and British close air support and interdiction (the British had reestablished air superiority) were key in Rommel's repulse. Rommel was severely defeated near Alam Halfa Ridge losing men, material, tanks, and aircraft that could not be replaced largely because of Malta.[198]

Montgomery then paused for two months before attacking. One might wonder what would have occurred if Rommel had defied Hitler and retreated to prepared positions in his rear during this build-up. When Montgomery did attack he had a three to one advantage in men, artillery, tanks, and most of all aircraft.[199] During this time JG 27 and JG 53, the two German fighter wings in North Africa were all but destroyed as fighting forces.[200] After prying permission to withdraw from Hitler, Rommel abandoned most of his Italian units and began a 1400 mile retreat. Two weeks later, the Allies landed in Algeria and Morocco. The race for Tunisia was on. The tide in North Africa had turned.

Rommel's successful 1,400 mile retreat from El Alamein to Tunisia should never have occurred. The British had air superiority all during the fight at El Alamein and British fighters and bombers should have shot this retreating army to pieces. It was a failure of tactical airpower that allowed Rommel to retreat intact. Nor did Montgomery pursue Rommel aggressively.

New Combined Anglo-American leaders were appointed under General Eisenhower when Montgomery entered Tunisia. Air Marshall Sir Arthur Tedder became the commander of Allied Air forces and pushed through the necessary Allied air-to-ground coordination.[201] It was no accident that American Tactical Air Doctrine: Centralized control, fighter-bombers rather than attack aircraft, air superiority first, then interdiction and close support, was all first codified during this time. Air to ground radio communications were standardized including frequencies, channels, terminologies, and a phonetic alphabet was adopted. It was the first attempt at coordinating large Anglo-American air and ground units. Judging by tac air's failure in the pursuit from Egypt, it was necessary.[202]

Hitler poured resources into Tunisia in 1943 that Rommel could have put to better use in 1942 in Cyrenaica. It was a lost cause and Hitler lost all of his forces including the belated reinforcements. The German surrender in Tunisia was almost as decisive in terms of men and material as Stalingrad. Almost 250,000 Axis troops surrendered with all of their equipment. Allied air power dominated the Med and never again would lose air superiority in the Mediterranean during the war.[203]

Using this command of the air and the lack of motivation of some Italians, the island of Pantelleria was bombed into surrender a few weeks later. On July 10, 1943, Sicily was invaded and was completely occupied by mid-August. The invasion of Italy followed with a secondary attack across the Straits of Messina on 3 September followed by the main attack just below Naples at Salerno on Italy's west coast on 9 September. The Germans were expecting a landing at Salerno as they had calculated the range of Allied fighters from Sicil-ian airbases and posted the few beaches suitable for landing. The 16th Panzer Division was waiting in the hills above the beaches at Salerno. Allied airpower was brought to bear as both sides attempted to reinforce. General Eisenhower requested and got strategic bomber support. General Ridgeway jumped the 82nd Airborne directly in support of the beaches as the Allied fighters protected the beaches from the Luft-waffe.[204] With naval gunfire support and close air support, the Allies consolidated the beachhead.

EASTERN EUROPE

When Hitler invaded Russia, the Germans achieved tactical and strategic surprise. Why this was true remains a mystery today. There was more than adequate intelligence to predict the attack. Stalin was given exact information from a number of sources on "Barbarossa" but none of this information was relayed to the VVS. And in spite of this the VVS Commander in Chief was summarily shot. If we can believe history, it was Stalin who should have been shot.[205]

The Luftwaffe dominated the battlefields of 1941. The well-trained German fighter pilots had technical and numerical superiority until the winter of 1941 when the weather made it difficult to operate. The Luftwaffe regained command of the air during the spring and summer of 1942 until the Russian counter-offensive at Stalingrad when the VVS flooded the battlefield with new high performance fighters.[206] The huge production of the USSR had caught up with the Germans and air superiority/supremacy belonged to the VVS from 1943 for the rest of the war.[207] Marshall Novikov had 17 air armies by May 1942. With no Strategic Missions (less than 5% of the resources were dedicated to strategic bombers), the Russians saw the VVS as long range artillery. They formed their fighters, attack aircraft, light and medium bombers into air armies and attached them to Russian Army Groups (Fronts). They also placed air armies in a general mobile reserve to be switched anywhere along the battlefield front. By 1943, a large amount of VVS aircraft were in these reserves and could be switched from one offensive area to another very quickly to dominate the battlefields.[208]

In the summer of 1943, Hitler had to have a victory on the Russian front. Only by the genius of Field Marshall Erich von Manstein had the Germans survived the winter of 1942-43 after the collapse at Stalingrad. Hitler decided to attack the salient near Kursk. Alerted through the Allies (Ultra) and Hitler's poor security, the Russians built an enormous defensive trap of fortifications, mines, and massive armor and infantry resources.[209]

Included with this Russia Front of Armies was an Air Army of about 2,900 (500 day bombers, 400 night bombers, 1,060 fighters, and 940 ground attack) aircraft with additional reserve air armies standing by. By stripping the entire Russian front, the Germans massed 2,050 (1,200 bombers, 600 fighters, 100 ground attack, and 150 reconnais-

sance) aircraft. This lineup gave the Soviets three times as many fighters and ground attack aircraft as the Germans.[210]

The Battle of Kursk was the massive anti-climax to Stalingrad for the Germans. It was the biggest armor battle in history and a bloody defeat for the Germans. They were never again able to challenge the Soviets on the Strategic Level. In the air, the Germans would never again have any semblance of numerical parity and the Russian dominance progressed steadily from air superiority to air supremacy. For the Germans, the war in the East was lost. It was just a matter of time.[211]

THE BATTLE FOR GERMANY

If Kursk was the biggest tank battle in history, the biggest air battle in history began in 1942 when U.S. strategic bombers arrived in England to begin daylight precision attacks into Germany. The tenets of Douhet, Trenchard and Mitchell were about to be tested. Brigadier General Ira C. Eaker commanded the initial force and was head of the Eighth Bomber Command. The Eighth Air Force Commander was Carl "Tooey" Spaatz.

The first American heavy bomber raid was flown on August 17, 1942 against the rail yards at Rouen in France. Twelve B-17s were escorted by Spitfire IXs of the RAF. The Spitfires kept the German Me-109s and FW-190s away from the bombers. Two Spitfires were lost as were two Me-109s. Brig. Gen. Eaker flew as an observer in the lead aircraft in the second element of bombers. The

raid was commanded by Colonel Frank Armstrong (the hero of the film "12 O'Clock High), and his co-pilot was Major Paul Tibbetts who three years later would pilot the Enola Gay over Hiroshima.[212]

When the Americans first landed in England, Gen. Eaker, when asked to speak to a British gathering, gave a very short two-line speech that set the tone for the American Eighth Air Force in England for the entire war:

"We won't do much talking until we've done more fighting. After we're gone, I hope you'll be glad we came." [213]

Just six weeks after arriving, Eaker met with General Arnold, Chief of Staff of the Army Air Force and General Marshall, Chief of Staff of the Army. When discussing the invasion of Europe from England and its requirements, Marshall commented, "I don't believe we'll ever successfully invade the continent and expose that great armada unless we first defeat the Luftwaffe".

Eaker replied:

"The prime purpose of our operations over here, aside from reducing their munitions capability, is to make the Luftwaffe come up and fight. If you will support the bomber offensive, I guarantee that the Luftwaffe will not prevent the cross-Channel invasion".[214] Both men lived up to their bargain.

A very persistent high level argument took place between the Americans and British about daylight precision bombing. The Americans still believed that they could penetrate German defenses in heavy bombers without fighter escort and without prohibitive losses. The British were the first to admit that round-the-clock bombing would be very useful, but from their experiences they did not believe the Americans could survive the German day fighters and Flak. The British also did not believe that a long-range fighter could be produced that could escort the bombers into Germany and still hold its own against the existing Luftwaffe fighters. The dispute was elevated to Churchill and Roosevelt at the Casablanca summit meeting in Morrocco. There Churchill talked to Eaker and subsequently decided to withdraw his objections to daylight bombing. The directive issued by the Combined Chiefs stated the following:"You should take every opportunity to attack Germany by day, to destroy objectives that are unsuitable for night attack, to sustain continuous pressure on German morale, to impose heavy losses on the German fighter day force and to contain German strength away from the Russian and Mediterranean theaters".[215]

The first strike into Germany was flown on January 27, 1943, the target Wilhelmshaven. The bombers claimed 22 Luftwaffe fighters shot down when the Germans actually lost seven. The raid was made with 91 bombers; 55 made effective runs on targets; one B-17 and two B-24s were lost.[216] Fighter escort was effective as far as it could go. Spitfires Vs, then the better Spitfire IXs, accompanied the bombers across the channel, then had to turn back and available P-38s went on as far as

they could go.

VIII Fighter Command had been scheduled for three groups of P-38s that were siphoned off to North Africa. The P-38 had the range to accompany the bombers until they dropped their external tanks. With their limited internal fuel the P-38s had to turn for home almost immediately once they dropped their tanks.[217] In addition, the P-38s were having a great deal of trouble with their Allison engines at altitude in Northern Europe. Later in the spring, the remaining P-38s were assisted by the P-47 which was initially limited by radio problems (the "Jugs" could talk to no one and radios had to be completely changed).[218] But at best the P-47's range was limited until pressurized external drop tanks could be developed for it later in July 1943. The P-47s flew their first combat mission on April 16, 1943 and held their own with the Luftwaffe's best FW-190s at altitude. It proved to be a viable fighter in the ETO.[219]

However, it quickly became obvious that though it was more difficult for the Luftwaffe to disrupt the American heavies with their high altitude formations, the British were right: unescorted bombers were savaged by the Germans. As soon as the fighter escorts turned back, the Germans struck with their fighters throwing everything at the American bombers, but the proverbial "kitchen sink". The Luftwaffe quickly found the bomber boxes were weakest defensively on the nose and began to attack from "Twelve O'clock High" – diving through from a straight ahead position. They loaded down their fighters with heavy cannon, mortars, and rockets that could fire at the bomber boxes

out of range of the B-17s/B-24s' .50 caliber guns.[220] They even tried to drop bombs on the bombers. There were several recorded successes.[221]

The summer period of 1943 was called "the Bloody Summer" as the VIII Bomber Command continued to bomb in spite of heavy losses of about 8.5%. It took no rocket scientist to point out that after 13 missions the entire force would be lost. No bomber crews in the Eighth worried about chest colds that summer. Life appeared far too short to "sweat the small things".[222]

In July, incredible air battles took place. On the night of July 24, Britain, who was fulfilling the night goals of around the clock bombing, chose to use one of her secret weapons, "Window". British engineers had discovered that aluminum foil cut in strips to match radar frequencies could be dropped into the air and totally clutter enemy radar scopes. They called these strips "Window". As the British bomber stream began its ingress into the continent (the British bombers attacked singly in long bomber streams), all of the German radars were blotted out with the clutter of thousands of targets as the British used "Window" (called Chaff today) for the first time. The entire German defensive system broke down chasing 11,000 apparently in-coming bombers as each strip of chaff appeared on radar as a bomber. 940 real British bombers crossed the target dropping 2,396 tons of bombs, 940 of them incendiaries.[223] The USAAF hit Hamburg the following two days and the RAF again on 27-28 July. The night of July 27-

28, 729 RAF bombers dropped an additional 1,200 tons of bombs and incendiaries on Hamburg. With fire fighting services exhausted, for the first time a man-made firestorm was created (to be duplicated in Tokyo and Dresden later). 50,000 Germans perished in the tornadic winds of fire that swept Hamburg. 900,000 people were left homeless.[224]

Albert Speer told Hitler that six more raids like Hamburg would end the war. After Hamburg was hit again on the 28 July, Goering switched German aircraft production priorities to fighter production.[225] The Germans transferred several veteran fighter Geschwader from Russia to Germany to meet the assaults. The air campaign in Germany guaranteed Russian air superiority on the Eastern Front.[226]

During July, the daylight campaign continued and the USAAF persisted in its deep unescorted strikes into Germany with from 300-350 bombers available at any one time. During this period, the USAAF lost 128 bombers and 1,200 crewmen while the Germans lost 40 fighters. In addition, 25-40% of all aircraft returning from Germany were damaged in battle.[227]

Realizing the bomber's vulnerability, the USAAF escorted them as far as fighter ranges permitted. By July 1943, when the P-47 received pressurized external tanks, the bombers were escorted to the German border before the P-47s had to turn back. At first, the German fighter pilots were wary of the P-47s which with their new drop tanks were escorting the bombers

as far as the German border. But the "Jugs" stuck close to the bombers, following their orders from VIII Fighter Command, escorting in close formation as the Luftwaffe had done in 1940 over England.[228]

Remembering well, Galland recommended that the German fighters should attack the American fighters while they were at a disadvantage and were tied to the bombers (as the Spitfires had done to him in 1940). The Luftwaffe high command ignored Galland and insisted on bomber attack. The result was deadly for the Germans as the P-47s assumed the offensive while the Germans attacked the bombers. The German fighters would attack from twelve o'clock and split-s and dive away from the heavy bomber defensive fire (they were used to diving away from Spitfires). Every American fighter could outdive the Luftwaffe fighters and nothing on earth could dive away from the "Jug". The Americans overhauled the Germans and shot them to pieces, particularly the Me-109 that had been structurally limited to about 375 MPH.[229]

The USAAF continued its deep penetrations into Germany unescorted while seeking an answer to long-range fighter escort. From July 24 to Aug 17, 1943 the USAAF hit a myriad of targets. During the first week the Eighth flew 1,720 sorties, 1,000 reaching the targets, 88 aircraft were lost. Targets included factories in Norway, Kiel, Hannover, Hamburg, Kassel, and Warnemunde. Many of the aircraft that returned were too damaged to fly again without extensive repairs. VIII Bomber Command was whittled down to only 200

bombers. Losses averaged 8.5%.[230]

On August 17, VIII Bomber Command produced a maximum effort against Regensburg and Schweinfurt. Brigadier General Curtis LeMay led the Regensburg raid of 146 with 127 aircraft reaching the target. It obliterated the Messerschmitt factory there setting back Me-109 production for five months. More importantly, the fuselage jigs for the Me-262 were destroyed. The USAAF lost 24 aircraft, almost 19%.

General Robert B. Williams' force went after the Schweinfurt ball-bearing factory with 230 aircraft. Because of poor USAAF fighter and bomber coordination, there was little or no fighter escort. The Schweinfurt force subsequently was attacked from Belgium all the way to the target and all the way out. 183 bombed the target, and 36 were lost, a 19.7% loss rate. Ball-bearing production dropped from 140 tons to fifty tons, but shortages were made up from Sweden and Albert Speer immediately undertook a ball-bearing production dispersal program.[231]

In these two raids, the USAAF lost 60 aircraft, 30 had to be left in Africa (LeMay had shuttled to North Africa) and 28 that got back to England were heavily battered. It was a victory for the Luftwaffe. Hitler did not see it that way and severely upbraided Hans Jeschonnek, the Luftwaffe Chief Of the General Staff for failing to stop the raid.[232]

LUFTWAFFE NIGHT FIGHTER OPERATIONS

That same night, after a brilliant and

successful night attack by the British on Peenemunde, Jeschonnek, who had predicted the Peenemunde attack would hit Berlin, committed suicide. He left a note very much like Udet. To be caught between Hitler and Goering was to cease to be.[233]

Jeschonnek had thought the raiders would hit Berlin and the new "Wild Boar" (Wild Sau) fighters had flown through and held in Berlin's flak alley for an interminable time sustaining some losses from "friendly" flak. The Wild Boar pilots were ordinary single-engined fighters used as night fighters without radar guidance. They entered the bomber stream by following the directions of shadowing Luftwaffe aircraft and braving their own flak. In August 1943, they were initially successful. They were to founder terribly in the horrible weather of the winter of 1943-44. Operational losses in the winter weather were horrendous. Fighter General Galland estimated this experiment cost him over 1,000 experienced pilots in a Luftwaffe dramatically short of experienced pilots. Wild Sau was discontinued in 1944.[234]

One of the untold stories of WW II was the build-up of German night and all-weather air defenses. At the time of the Battle of Britain night fighter defenses were non-existent. German night defense was almost totally dependent on flak. To organize an air-to-air night fighting defense, the required bands of radar, operations centers, command and control communications, and radar-equipped night fighters had to be built from scratch. The construction of these networks was initially the responsibility of General Josef Kammhuber. Kammhuber was a good administrator but demanding and inflexible. He did not get along with Galland or Goering, but the system he built under the stress of war was functional. By 1943-44 it had cost the British thousands of combat casualties. For example, the British flew 20,224 sorties between Nov. 1943 – March 1944. Of these 9,111 were against Berlin, 2,700 were shot down or heavily damaged, 1,047 were lost over Germany. This system would not have been in being except for Kammhuber. Kammhuber was sacked by Hitler for believing actual U.S. production figures that Hitler refused to accept. He was relieved on June 24, 1943 for arguing with Hitler.[235]

When the Allies struck Hamburg on July 24, 1943, Kammhuber's system failed because of British methods and chaff. The firestorm in Hamburg united Germany (particularly the Luftwaffe) in terms of air defense as never before.[236] It, however, doomed Kammhuber's association with the night defense effort. He was relieved and the responsibility given to an already overburdened Galland.[237]

The problem with Kammhuber's system was the zones he set up could handle only single attackers. When the British blinded the German radars with chaff and attacked in small formation/bomber streams, the system broke down. His sector system for fighters, flak, etc. was not sufficently flexible to handle a large number of bombers that fast.[238]

The best of the German night fighters was the Heinkel 219, "Owl", which was designed and built as a night fighter. The Ju-88 and Me-110 were also used. Hitler and Milch preferred the later as these aircraft were already in production and they did not wish to divert resources to a third aircraft.[239] Heinkel seemed to inherit this refusal to produce additional types of aircraft (remember the He-100?).

THE DAYLIGHT WAR

The losses on the first Regensburg and Schweinfurt raids had hurt the USAAF. The VIII Bomber Command did not fly into Germany again until September 6, 1943 when it sustained 14.4% losses attacking Stuttgart.[240]

If proof was needed that the unescorted policy was flawed, it came with a vengeance when on October 18, 291 bombers went back to Schweinfurt. Two-hundred-twenty-nine reached the target, 60 aircraft were shot down, and 121 more heavily damaged. The loss rate was 26.5% which no air force could sustain. The Luftwaffe was winning the war for air superiority.[241]

Another problem for VIII Fighter Command was Brig. Gen. "Monk" Hunter's close escort doctrine. Like Goering in the Battle of Britain, he insisted the fighters closely escort the bombers.[242] This was pure Army Tactical School doctrine and totally wrong. The young American fighter leaders, like Adolph Galland before them, protested mightily against this tactical myopia. Hunter was relieved later by Arnold even though they were old friends dating back to WW I. Unfortunately, Hunter's escort doctrine survived his relief.

General Arnold had two heart attacks during this time and was very impatient with VIII Air Force and General Eaker. He did not seem to understand the difficulties of building an air force under the combat conditions in the European Theater Of Operations (ETO). Arnold had made many statements about the B-17 Flying Fortress and its capabilities, most based on his experiences in the past. Robert A. Lovett, the brilliant Assistant Secretary for Air, visited Eaker and believed that Arnold blamed Eaker for the failure of the B-17 to perform in combat the way that he had said it would. After his visit to Eaker, Lovett pushed hard on Arnold to supply long-range escort fighters for the bombers. Arnold reverted to his 1930s beliefs and stated to Lovett that B-17s were our only need, that's all; very few fighters can keep up with them. This, when Me-109s and FW-190s were having no trouble catching them and lobbing rockets, mortar rounds and large caliber cannon shells into the massed formations.[243] Arnold was wrong and he knew it.

The fight between Eaker and Arnold was unfortunate and bitter. Eaker was "kicked upstairs" to the Med in December of 1943 after he had done an incredible job. It was not Eaker's fault that Arnold had not grasped until 1939 what most other air force leaders in the world had already learned from WW I, Spain, China and Manchuria: unescorted bombers were dead meat.[244] The USAAC had started too late to build an escort fighter. Eaker knew that long range

fighters were necessary , but with none available, he had the heart-wrenching job of launching heavy bombers into Germany day-after-day. He had to face the crews and their terrible losses. Probably for this reason, he continued to espouse that heavy bombers could get through unescorted. (What other choice did he have until he was furnished with long range fighters?) Eaker fought it as much as he could, but in the end saluted, and followed his orders.[245] As the overall air force commander in the Med, he continued his stellar leadership and because of his leadership the answer to the problem was near at hand.

Lovett's push on Arnold on behalf of Eaker was to have some positive results. Lovett insisted that the answer was immediate long-range fighter escort. Arnold told Lovett that the P-38 was the only aircraft in production that could provide such escort, but he thought it was too slow. Lovett pointed out that with its low internal fuel it had to turn back very soon after its external tanks were jettisoned. He also pointed out that he had high hopes for the P-51 with wing tanks.[246]

Arnold sent his Chief of Staff, General Barney Giles, to see J. H. "Dutch" Kindelberger, the president of North American Aviation. Tommy Hitchcock, the great polo player and long-time friend of Arnold's, had suggested a switch from the Allison to the Rolls-Royce Merlin with its two-stage supercharger to improve the Mustang's altitude performance. Tests quickly showed the P-51/Merlin to be superior to every other fighter in the world. The key to its

fantastic performance was its NACA laminar flow wing, low-drag design and the additional power of the Merlin engine.[247]

But it did not have the legs for long range escort once external fuel was jettisoned. Giles suggested removing the radio behind the cockpit installing a 100 gallon tank and squeezing in 100 gallons in each wing. Kindelberger did not believe the landing gear and structure was strong enough to take the additional 300 gallon load, but tests revealed that it was.[248]

Giles also visited Lockheed and that company squeezed in 120 more gallons into the wings of the P-38J. By mid-1943, the USAAF had two fighters that could reach the deepest targets inside Germany.[249]

The first long-legged P-51Bs arrived in England in November 1943 and first flew in December 1943 too late to save Eaker. Incidentally, Eaker took on historians who stated after the war that the failure to plan and build long range fighters prior to the war was inexcusable. He said everyone including Arnold would have done anything for a long-ranged fighter. Eaker pointed out that it was just beyond the state of the art to build such a fighter until the P-51 showed up. In fact, the USAAF had experimented with the P-38 extensively, but its internal fuel capacity was too low.[250] Nor was the P-38 respected as an air-to-air fighter in Europe due to some serious engine problems.[251] The Germans thought it was easy meat.[252]

Judging from British and German reaction to the P-51, they thought the low-fuel Spitfire and FW-190 were the pinnacle of design and could not believe that the P-51 was outmaneuvering Me-109s and FW-190s over Berlin. As already pointed out, this long range fighter problem was fundamental to the British adopting night area bombing and believing the Americans should adopt night bombing too. Goering had assured Hitler that such a fighter was impossible. Hitler refused to discuss planning for long range escort based on Goering's ignorance. This was incredible. Wonder why no one paid any attention to Japanese operations in China where a fighter called the Zero was routinely operating over ranges in excess of the Germany runs. The Japanese were the Germans' allies. Perhaps we should have sent them Chennault's report. Nobody else read it! British observers sent much the same information on the Zero to London in 1940.

TRAINING

One item largely ignored in most histories is the importance of USAAC/USAAF training programs during WW II. General "Hap" Arnold was the architect of this training program. Unlike every other air force in WW II, the U.S. was never short of pilots and our fighter pilots were the best-trained in the world. Not only did our fighter pilots get more flying time in training than others, but they flew better eqipment. The AT-6 (later T-6) built by North American simulated modern WW II fighters very well. Over 20,000 were built and were extensively

used by our Allies as well. But the most important training adjunct for war in the European Theater was the U.S. emphasis on instrument training. This emphasis was originally recommended by Jimmy Doolittle to General Arnold. After some experience in Europe, a formal USAAF instrument school was established in 1942. Instrument training was incorporated into the basic training program in 1943. Hap Arnold's role in building the overall USAAF pilot training program alone have assured his place in history as one of America's greatest air leaders.

Conversely, the Luftwaffe did not emphasize instrument proficiency to the same extent and encountered shortages throughout the war. They were often unable to operate in bad weather. For example, when the Luftwaffe needed night fighter pilots, German leaders had difficulty finding enough good instrument pilots. When the large air battles of 1943-1944 began and losses began to mount, the new day fighter pilots had inadequate instrument training. Operational losses sometimes exceeded combat losses because of bad weather. Too late, Goering instituted a "crash" instrument training program during 1944.

SUMMARY

The years 1941-1943 were the most conclusive of the war. The first two years were all Axis. The Balkans, North Africa, Russia, Pearl Harbor, Malaysia, the Philippines, the Dutch East Indies, all of the Central Pacific – all were victories for the bad guys. The Luftwaffe and Imperial Japanese Air Force reigned supreme until mid-

to-late 1942. The AVG proved that the Japanese were not invincible and the U.S. Navy confirmed it at the Coral Sea and Midway. In Russia, the Germans were defeated at Stalingrad and the Migs, Laggs, and Yaks wrested air superiority from the Germans over the battlefield. In the Mediterranean, the British fought an all-out air battle for Malta and in the end won (the Luftwaffe was stretched thin between Russia, Germany, and the Med). El Alamein followed with British control of the air and Rommel's final defeat.

The American invasion of French North Africa forecast the beginning of the end. German defeat in Tunisia and the invasions of Sicily and Italy followed. The Strategic Air Offensive started around-the clock against Germany.

In 1943, the tide turned for the Allies. Italy surrendered. Victories at Guadalcanal, New Guinea, Kursk, Sicily, and Italy followed. And allied air power had gained air superiority everywhere except over Germany. By the grace of God, the Germans had not realized that the Me-262 was the answer to their air battle prayers. As the year drew to a close, the issue of command of the air over Germany was still in doubt, but any educated observer could see that the war was over. The production lines of the U.S., USSR, and Great Britain were in full-stride. It was just a matter of time.

NOTES

1. Benjamin S. Kelsey, *The Dragon's Teeth?*, Smithsonian Institution Press, Washington D.C., 1982, pp. 46-47.
2. Ibid.
3. Duane Schulz, *The Maverick War Chennault And The Flying Tigers*, St. Martin's Press, New York, 1987, p. 17: Quote by General Bruce Holloway.
4. Edward H. Sims, *Fighter Tactics And Strategy 1914-1970*, Harpers & Row, New York, 1972, Part Three, p. 87.
5. Op.Cit.; p.89.
6. Alfred Price, *Fighter Aircraft*, Arms and Armour Press, London, 1990, pp. 8-16.
7. Ibid.
8. Ibid.
9. Ibid.
10. William Green and Gordon Swanborough, *The Complete Book Of Fighters*, Smithmark Publishers, Salamander Books, London, 1994, p. 261.
11. Op. Cit.; pp. 476-477.
12. Op. Cit,, pp. 425-426.
13. Op. Cit.; pp. 405-406.
14. Op. Cit.; pp. 358-359.
15. Op. Cit.; pp. 485-486.
16. Green, Op. Cit.; p. 408.
17. Op, Cit.; pp. 409-410.
18. Green, Op. Cit.; p. 408.
19. Walter J. Boyne, *Clash Of Wings*, Simon & Schuster, New York,1994, pp. 36-38.
20. Price, Ibid.
21. Alfred Price, *Fighter Aircraft Combat Development In World War II.*, Arms And Armour Press, London, 1976, p. 78.
22. Dewitt S. Copp, *A Few Great Captains*, Doubleday & Co., New York, 1980, p. 457.
23. Caidin, Op. Cit.; p. 292.
24. Boyne, Op. Cit.; p. 42.
25. Boyne, Op. Cit.; p. 208.
26. Richard Hallion, *The Rise Of The Fighter 1914-1918*, Nautical & Aviation Publishing Co., Baltimore, Maryland, pp. 151-152.
27. Len Deighton, *The Fighter: The True Story Of The Battle Of Britain*, Jonathan Cape, London, 1977, pp. 126-130.
28. Duane Schultz, *Maverick War, Chennault And The Flying Tigers*, St. Martin's Press, New York, 1987, p. 62.
29. Boyne, Op. Cit.; p. 22.
30. R.J. Overy, *The Air War 1939-1945*, Stein & Day Publishers, New York, 1980, p. 11.
31. Ibid.
32. Overy, Op. Cit.; pp. 35-36.
33. Ibid.
34. Boyne, Op.Cit.; pp.28-42.
35. Ibid.
36. Ibid.
37. Ibid.
38. Martin, Gilbert, *The Second World War*, Henry Holt & Co., New York, 1989, p. 45. The Germans had broken the British Naval Code.
39. Vincent J. Esposito, *The West Point Atlas Of American Wars: Vol II*, Frederick A. Praeger, New York, 1959, Map 11.
40. Sims, Op. Cit.; pp.114-117.
41. Ibid.
42. Ibid.
43. Green, Op Cit.; p. 409.
44. Green, Op. Cit.; p. 71.
45. Johnson, Op. Cit.; p. 132.
46. Gilbert Martin, *The Second World War*, Henry Holt & Co., New York, 1989, pp. 62-63.
47. Esposito, Op. Cit.; Map 12.
48. Ibid.
49. Gilbert, Op. Cit.; p. 65.
50. Esposito; Ibid.
51. Esposito, Op. Cit.; Map 15.
52. Gilbert, Op. Cit.; pp. 75.

53. Esposito, Ibid.
54. Esposito, Op. Cit.; Map 17.
55. Thomas E. Griess, *The Second World War, European and Mediterranean Theater: West Point Military History Series*, Avery Press, Wayne, New Jersey, 1989, p. 73.
56. Ibid.
57. Deighton, Op. Cit.; pp. 285-286.
58. Griess, Op. Cit.; pp. 68-69.
.59 Deighton, Op. Cit. p. 151.
60. Deighton, Op. Cit.; p. 134.
61. Overy, Op. Cit.; p. 41.
62. Overy, Op. Cit.; p. 64.
63. Op. Cit.; p. 150.
64. Op. Cit.; p.41.
65. Sims, Op. Cit.; pp. 89-90.
66. Op. Cit.; p. 113.
67. Op. Cit.; p. 225.
68. J.E. Johnson, *The Story Of Air Fighting*, Bantam Books, New York, 1985, p. 117.
69. Deighton, Op. Cit.; pp. 138-139.
70. Werner Baumbach, *The Life And Death of the Luftwaffe*, Ballantine Books, New York, 1960, pp. 79-87.
71. Johnson, Op. Cit.; pp. 100-101.
72. Op. Cit.; pp. 117.
73. Griess, Op. Cit.; p. 77.
74. Griess, Op. Cit.; pp. 76-77.
75. Deighton, Op. Cit.; p. 157.
76. Ralph Barker, *The Epic of Flight The RAF At War*, Time-Life Books, Alexandria, Va., 1981, p. 49.
77. Griess, Op. Cit.; p. 75.
78. Alexander McKee, *Strike From The Sky*, Lancer Books, New York, 1960, p. 31.
79. McKee, Op. Cit.; p. 31.
80. Griess, Op. Cit.; p. 75.
81. Alfred Price, *Battle Of Britain:The Hardest Day*, Charles Scribner Sons, New York, 1979, p. 13.
82. Raymond E. Toliver & Trevor J. Constable, *Fighter General The Life Of Adolph Galland*, AmPress, Zephyr, Nevada, 1990, p. 93.
83. Op. Cit.; p. 7.
84. McKee, Op. Cit; p. 76.
85. Deighton, Op.Cit.; p. 199.
86. Barker, Op. Cit.; p. 58.
87. Deighton, Op.Cit.; pp. 199-203.
88. Op. Cit.; pp. 204-206.
89. Deighton, Op. Cit.; pp. 223. *Maximum age for Geschwader Commanders 32, for Gruppe 30, for Staffel 27.*
90. Barker, Op Cit; pp. 62-63.
91. Deighton, Op. Cit.; p. 230.
92. Tolliver, Op. Cit.; p. 108.
93. Barker, Op. Cit.; p. 67.
94. Ibid.
95. Price, Op. Cit.; p. 169.
96. Griess, Op. Cit.; pp. 75-76.
97. Price, Op. Cit.; pp. 50-51.
98. Johnson, Op. Cit.; p. 159.
99. Deighton, Op. Cit; p. 151.
100. Eugene M. Emme, *The Impact Of Air Power: "Defeat Of The Luftwaffe"*, by Adolph Galland; Van Nostrand; New York; 1959; pp. 245-255.
101. Griess, Op. Cit.; p. 83.
102. Price, Op. Cit.; p. 165.
103. Emme/Galland, Ibid.
104. Ibid.
105. Green, Op. Cit.; pp. 297-298.
106. Deighton, Op. Cit.; p. 223.

107. Schultz, Op. Cit.; p. 75.
108. Green and Swanborough, Op. Cit.; p. 409. The Zero had a max speed of 331 MPH at 15,000 feet, could outclimb and out-turn any other contemporary fighter, had two 7.9 MM guns and two 20 MM low velocity wing cannon, with a range of 1930 miles, and was carrier qualified.
109. Schultz, Ibid.
110. Griess, Op. Cit.; pp. 93-94.
111. Ibid.
112. Ibid.
113. Griess, Op. Cit.; pp. 92-101.
114. Ibid.
115. Griess, Op. Cit.; p. 107.
116. Ibid.
117. Op. Cit.; p. 134.
118. Sims, Op. Cit.; pp. 120-127.
119. Ibid.
120. Ibid.
121. Christopher Shores, *Fighter Aces*, Hamlyn Publishing Group, London, 1975, p.107.
122. Donald L. Caldwell, *JG 26 Top Guns of the Luftwaffe*, Ivy Books, New York, 1991, pp. 94-95.
123. Ibid.
124. Overy, Op. Cit.; pp. 110-116.
125. Green, *The Complete Book Of Fighters*, Op. Cit.; pp. 559-560.
126. Op. Cit., pp. 286-287.
127. Op. Cit.; pp. 214-216.
128. William Green and Gordon Swanborough, *Soviet Air Force Fighters Part Two*, Arco Publishing Company, New York, 1978, pp. 50-58.
129. William Green, *Fighters, Volume Two*, Hanover House, Garden City, New York, 1961, pp. 163-165.
130. Green, Comp. Encycl., Op. Cit.; p. 426.
131. Op. Cit.; p. 409.
132. Enzio Angelucci, *The American Fighter*, Orion Books, New York, 1985, p. 128.
133. Op. Cit.; p. 113.
134. Deighton, Op. Cit.; pp. 292-293.
135. Ibid.
136. Charles W. Cain, *Fighters of World War II*, Exeter Books, New York, 1979, pp. 22-27.
137. Ibid.
138. Green & Swanborough, The Complete Book of Fighters.
139. Copp, Op. Cit.; pp. 455-457.
140. Overy, Op. Cit.; Table pp. 192-193.
141. Deighton, Op. Cit.; p.134.
142. Cajus Bekker, *The Luftwaffe War Diaries*, Da Capo, New York, 1994, p. 313.
143. Op. Cit.; p. 197.
144. Bekker, Op. Cit.; p. 350.
145. Overy, Op. Cit.; Table pp. 192-193.
146. Op. Cit.; p. 198.
147. Ibid.
148. Robin Cross, Citadel *The Battle Of Kursk*, Michael O'Mara Books, London, 1993.
149. Overy, Op.Cit.; p. 71.
150. Cross, Op. Cit.; p. 110.
151. Ibid.
152. Cain, Op. Cit.; pp. 120-128.
153. Deighton, OP. Cit.; p. 134.
154. Overy, Op. Cit.; Table pp. 192-193.
155. Ibid.
156. Overy, Op. Cit.; pp. 193-198.
157. Overy, Op. Cit.; Table pp. 192-193.
158. Ibid.
159. Boyne, Op. Cit.; p. 102.
160. Caidin, Op. Cit.; pp. 116-117.

161. Caidin, Op. Cit.; pp. 112.
162. Caidin, Op. Cit.; p. 117.
163. Thomas E. Griess, *World War II, Asia and the Pacific*, Avery Publishing Group, Wayne, New Jersey, 1989, p. 48.
164. John B. Lundstrom, *The First Team*, Naval Institute Press, Annapolis, MD., 1984, pp. 441-444.
165. Ibid.
166. Duane Schultz, *The Maverick War Chennault And The Flying Tigers*, St.Martin's Press, New York, 1987 p. 282.
167. Griess, Op. Cit.; pp. 74-75.
168. Louis H. Brereton, *The Brereton Diaries*, William Morrow and Company, New York, 1946.
169. Thomas G. Miller, *The Cactus Air Force*, Bantam Books, New York, 1969, p. XV.
170. Op. Cit.; pp. 209-213.
171. Op. Cit.; pp. xii-xiii.
172. Ibid.
173. Boyne, Op. Cit.; pp. 238-239.
174. Miller, Op. Cit.; pp. xii-xiii.
175. Green and Swanborough, Op. Cit.; pp. 341-343, 264-265, 586-587.
176. Boyne, Op. Cit.; pp. 234-239.
177. Ibid.
178. Ibid.
179. George C. Kenney, *General Kenney Reports*, Duell, Sloan & Pierce, New York, 1949, p. 127.
180. Op. Cit.; pp. 112-113.
181. *Douglas MacArthur, Reminiscences*, McGraw Hill, New York, 1964, p. 168.
182. Martin Caidin, *Fork Tailed Devil: P-38*, Ballantine Books, New York, 1971, Introduction.
183. Griess, Op.Cit.; pp. 137-138.
184. F. Dodson Stamps & Vincent Esposito, *Summaries Of Selected Military Campaigns*, Department of Military Art and Engineering, West Point, New York, 1952, p. 154.
185. Ibid.
186. Daniel E. Barbey, *MacArthur's Amphibious Navy*, Naval Institute Press, Annapolis, Md, 1969, p. 100.
187. Op. Cit.; p. 57.
188. Stamps & Esposito, Op. Cit.; p. 120.
189. Griess, Europe and Med., pp. 166-170.
190. Op. Cit.; p. 166.
191. Bekker, Op. Cit.; p. 236.
192. Op. Cit.; p. 245.
193. Op. Cit.; p. 239.
194. Op. Cit.; p. 243.
195. Op. Cit.; p. 244.
196. Op. Cit.; pp 246-253.
197. Stamps and Esposito, Op. Cit.; p. 121.
198. Griess, Op. Cit.; p. 170.
199. Ibid.
200. Galland, Op.Cit.; p.144.
201. Griess, Op.Cit.; p. 175.
202. Emme, Op. Cit.; pp. 237-244.
203. Stamps and Esposito, Op. Cit.; p. 124.
204. Griess, Op. Cit.; p. 234.
205. Griess, Op. Cit.; p. 107.
206. Cross, Op. Cit.; pp. 118-126.
207. Griess, Op. Cit.; pp. 121-135.
208. Cross, Ibid.
209. Ibid.
210. Ibid.
211. Op. Cit.; pp. viii.
212. Parton, "*Air Force Spoken Here*" General Ira C. Eaker & Command of the Air, Adler & Adler Publishers, Bethesda, Md.,1986, pp. 174-175.
213. Op. Cit.; p. 154.
214. Op. Cit.; p. 148.
215. Op. Cit.; p. 223.
216. Boyne, Op. Cit.; p. 305.

217. Parton, Op. Cit.; p. 279.
218. Op. Cit.; p. 265.
219. Op. Cit.; p. 265.
220. Boyne, Op. Cit.; p.315.
221. Bekker, Op. Cit.; pp. 316-322.
222. Boyne, Op. Cit.; p. 323.
223. Boyne, Op. Cit.; p. 318.
224. Boyne, Op. Cit.; p. 319.
225. Bekker, Op. Cit.; p. 313.
226. Op. Cit.; p. 319.
227. Boyne, Op.Cit; p. 312.
228. Op. Cit.; p. 228.
229. Ibid.
230. Boyne, Op. Cit.; pp. 323-324.
231. Op. Cit; pp. 324-325.
232. Ibid.
233. Bekker, Op. Cit; pp. 314-316.
234. Tolliver & Constable, Op. Cit.; pp. 237-238.
235. Boyne, Op. Cit.; p. 302.
236. Galland, Op. Cit.; p. 162.
237. Constable, Op. Cit.; p. 234.
238. Bekker, Op. Cit.; pp.215-216.
239. Boyne, Op. Cit.; p. 330.
240. Op. Cit.; pp. 326-327.
241. Ibid.
242. Parton, Op. Cit.; 264-265.
243. Op. Cit.; p. 279.
244. Stephen L. MacFarland and Wesley Phillips Newton, *Command Of The Sky*, Smithsonian Institutional Press, Washington, 1991, p. 35.
245. Boyne, Op. Cit.; p. 328.
246. Parton, Op. Cit.; p. 279.
247. Roger A. Freeman, *Mustang At War*, Doubleday And Company, Garden City, NY., 1974, p. 55.
248. Op.Cit.; p. 280.
249. MacFarland & Newton, OP. Cit.; p. 105.
250. Ibid.
251. Caidin, Op. Cit.; p. 185.
252. Bekker, Op. Cit.; p. 317.
253. Constable & Tolliver, Op Cit.; p 206.
254. MacFarland & Newton, OP. Cit.; pp. 71-80.
255. Ibid

Above: *The F6F Grumman Hellcat was designed specifically to out-perform the Mitsubishi AGM "Zeke" (Zero). With its 2000 hp Pratt Whitney engine, it was faster, better armed and armored than its opponent. It shot down more Japanese than any other U.S. Navy aircraft.*

CHAPTER 5

FIGHTER DEVELOPMENT IN WORLD WAR II

PART THREE 1944-1945

The Lineup – Allies Vs. Axis as of January 1, 1944

As before, it is useful to survey the status of fighter development at the beginning of the period to be examined:

• Britain - Spitfire IX, Rolls-Royce Merlin, 1,700 HP, 408 MPH, 2 x 20 MM, 2 x .50 caliber machine guns all wing guns. Improved version of the Battle of Britain fighter. [1]

• Royal Navy - F4U-1D, Corsair, R-2800 Pratt Whitney, 2,000 HP, 392 MPH, 6 x .50 caliber wing guns. Not accepted by U.S.Navy to land aboard ship until April 1944. British Navy used it aboard ship 9 months earlier than USN.[2]

• USAAF - P-51B, Mustang, Rolls Royce Merlin, 1,298 HP, 440 MPH, 6 x .50 caliber wing guns. Best long range fighter in the world. Twice the range of the Spitfire.[3]

• USN - F4U-1D, Corsair, R-2800 Pratt Whitney, 2,000 HP, 392 MPH, 6 x .50 caliber wing guns. Not accepted by U.S. Navy to land aboard ship until April 1944. British Navy used it aboard ship 9 months earlier than USN.[4]

• Germany - FW-190A, BMW 801D, 1,700 HP, 402 MPH, 4 x 20 MM cannon; 2 x 13 MM machine guns. Same basic fighter of 1943. [5]

• Japanese Army - KI-61, (TONY), Kawasaki Ha-40 Engine, 1,175 HP, 360 MPH, 2 x 20 MM. Only liquid-cooled engined Japanese fighter.[6]

• Japanese Navy - Type 0, (Zeke), Nakajima Sakae 21 engine, 950 HP, 331 MPH, 2 x 20 MM, 2 x 7.9 MM. Obsolete in 1944. Same basic Zero. [7]

• USSR - La-5, Shvetsov M-82F, 1,850 HP, 403 MPH, 2 x 20 MM Cannon, 2 x 7.6 MM wing machine guns. Some handling problems, but formidable at low altitude. [8]

Interestingly, the Allies showed marked improvement over Axis fighters. Look what the Me-262 could have done at this time with its 540 MPH top speed (it became operational in June 1944 but in small numbers). With the preponderance of Russian and U.S. production the Axis were already buried numbers wise.

After night operations got underway the night fighters also became very important:

Night Fighter Lineup -- 1944

• Britain -Mosquito Mk XIII, 2 x Rolls-Royce 1,300 HP, 370 MPH, 4 x 20 MM, 4 x .303 caliber machine guns, NF Mk XII radar.[9]

• USAAF - P-61, Black Widow, 2 x Pratt Whitney 2,250 HP, 4 x 20 MM 4 x .50 caliber, 362 MPH, SCR 10 Radar.[10]

• Germany - He-219, Uhu (OWL), 2 x DB603, 1,850 HP, 4 x 30 MM, 2 x 20 MM, 416 MPH, FG 220 Licenstein radar. [11]

If one had to pick a best fighter from this trio based on this data, it would have to be the Uhu. For some reason,

the Germans decided to only make 268 of them.[12] The Germans seemed to have had a penchant to back any design in major production regardless of its lack of quality and they refused to back other designs regardless of higher quality. An aircraft could not go into mass production unless it was already being mass produced. For a logical people this was ludicrous. Obviously someone making German production decisions had a hidden agenda.

The Battle For Germany

When the P-51s appeared over the target in January 1944 the Germans were shocked.[13] Unfortunately for the Allies, initially there were not enough P-51s and bomber losses remained high.[14] But with American high-performance fighters escorting bombers deep over Germany and their numbers increasing every day, it was just a matter of time. The American and British bombers went after the German fighter production and they were successful. The factories were heavily hit. In the Luftwaffe, there was genuine alarm.[15]

German interceptors, heavily-laden, single-engine fighters loaded down with cannon, rockets, mortars, etc. were "duck soup" for the American Mustangs as were all of the German two-engine "bomber shooters". The Mustangs were faster, more maneuverable and the heavily-laden aircraft had no chance. These "Zerstorers" had to be escorted by German single-engined fighters to survive; e.g., the II/ZG 26 Gruppen equipped with twin-engined Me-410A-2s carrying heavy 40 MM cannons had been very effective against American heavy bombers. On May 13, 1944, the unescorted Gruppen was caught by P-51s. The Gruppen was destroyed and never flew again as a unit.[16] The clumsy, rocket carrying single-seaters and Me-110s suffered similarly.[17] In February and March of 1944, the Luftwaffe lost over 4,000 aircraft. The staggering loss of trained aircrews was irreplaceable. The Luftwaffe no longer could meet every Allied raid. They were forced to pick and choose with less and less success.

During 1943, the USAAF made the same escort mistake the Germans had made during the Battle of Britain and over Malta. As directed by VIII Fighter command, they used the U.S. Army Air Tactical School tactic of closely escorting the bombers. But the young aggressive fighter leaders of VIII Fighter Command began clamoring to fight aggressively by attacking the German bomber shooters and fighters as they were forming up, before they could mass to go after the American bombers. This was the tactic

that Adolph Galland, the German fighter leader, feared the most.[18]

On January 6, 1944, General Kepner (who replaced General Hunter as VIII Fighter Commander) and the new Eighth Air Force Commander, General Jimmy Doolittle, turned them loose.[19] Doolittle gave the VIII Fighter Command the mission to destroy the Luftwaffe fighters in the air and on the ground, while VIII Bomber Command hit the factories that made them and the oil industry that fueled them. The USAAF fighter pilots responded with enthusiasm.[20] It was one of the best decisions of the war. Years later, General Doolittle thought it was his best. Galland, his opponent, agreed.[21]

The American fighters converged in massed groups all along the bomber path into Germany. Their mis-

sion was to attack the intercepting German fighters wherever they found them. Just as the Americans hoped, they often caught the German fighters as they were forming up and organizing to attack the bombers. Often Allied radio intercept personnel would direct the long range fighter-s to the organizing Germans. It was devastating. Then on the way home the Americans would drop down and attack the German airfields sometimes catching the Germans low on fuel and trying to recover, or catching them sitting on the ground.[22]

The USAAF fighters used the bombers as bait for the German fighters and attacked them everywhere they found them. With a seven-to-one advantage for the Allies, every German pilot was at risk everywhere, and they could not be replaced.[23]

By almost superhuman efforts, after the initial raids that appeared to destroy their fighter production, the Germans dispersed and rebuilt their aircraft production, and fighter production recovered rapidly. There were ample fighters, but no experienced combat pilots and no fuel.

The Luftwaffe was doomed in 1944 by the same problem that threatened the British with destruction in 1940: The lack of experienced aircrews; i.e., fighter pilots. Under incessant attack, the Luftwaffe lost the majority of its best pilots, and there was no way to recover. With odds of 7-1 the USAAF was crushing the Luftwaffe day by day, attacking everywhere. Air supremacy was nigh.[24]

In May, many of the heavy bombers were switched to tactical interdiction targets to make way for the invasion. With many medium bombers also attacking French transportation systems, the overall system was quickly reduced to a shell before the invasion. At the same time the Eighth reduced the German Synthetic Oil Industry to shambles and by September the entire system had broken down and reserves were gone. Total production met only 1/5 the Luftwaffe's requirements. American intelligence probably should have selected the oil targets earlier. [25]

The results are history. On D-DAY, 6 June 1944, there were only a few hundred operational Luftwaffe aircraft to oppose the thousands of the Allied Air Forces. With air supremacy, the Allied bombers ranged all over Europe. German reinforcements to Normandy had to run a veritable gauntlet of fire and arrived in battered decimated condition. The Allied air advantage in the odds quickly increased to 20-1.[26]

It was into this stacked deck that the Me-262 became operational, too little and too late. It was very similar to the situation with the Fokker D VII in WW I. The first Me-262s became operational in June 1944. Hitler had delayed the fighter production about four months, but production of the new turbine engines was the real roadblock. Over 1,400 Me-262s were deliverd to the Luftwaffe, but of those, only 300 were actually used in operations.[27]

Another radical interceptor was the Me-163 powered by an unreliable rocket engine. This little flying wing interceptor exceeded 600 MPH in 1941 and more than 700 MPH in 1944. The fuels and oxidizers were so volatile that they were sometimes more dangerous to the pilot than the enemy. Three hundred were built but achieved only nine kills with two probables. Its bat-like appearance and lightning speed were frightening to many Allied bomber crews. It is possible that it caused more casualties from fright than from actual combat.[28] It is also possible that German casualties from riding this rocket exceeded those caused by it in combat with the Allies.

Below Left: *The Vought F4U Corsair was the best performing Naval fighter of WWII.*

As the Allied Armies, in France, Italy, and Russia, pushed toward Germany their air supremacy increased into complete domination – the air war was over in Europe. Hitler should have surrendered in 1943 or in June of 1944 for sure. It was insane to continue. Many good men continued to serve that insanity to the last full measure. Others were its victims. All to no purpose.

The Vengeance Weapons

During 1944 Hitler's vaunted Vengeance Weapons, the V-1 and V-2, were unleashed against London. The V-1 was a small pilotless aircraft propelled by a ingenious pulse-jet engine. Top speed was about 400 MPH. Guidance was by a simple gyro sytem which held a heading. Range was controlled by a small prop-driven airlog that mea-

sured distance flown (similar to the system used on the Wright Flyer to measure distance). When the airlog measured the correct distance, the engine shut down and the missile nosed over to dive into the ground. The warhead was approximately 2,000 pounds of high explosive. It was accurate enough to hit a city as large as London or Antwerp.[29]

The V-2 was the first real ballistic missile. Its rocket motor used alcohol as fuel and liquid oxygen as an oxidizer. The rocket actually left the atmosphere and climbed to an altitude of about 60 miles and reentered following a ballistic path at hypersonic speed. Unlike the V-1 which could be intercepted by a fast fighter, or shot down by flak, there was no defense against the V-2 which also carried approximately 2,000 pounds in its warhead. Again it usually could hit a large city.[30]

More than 6700 V-1s crossed the English coast, 3900 were shot down. The remaining 2800 destroyed 23,000 homes and buildings and killed 5,500 civilians.[31]

Almost 6000 V-2s were built. 1054 fell in England and 1675 on the continent of which 1,265 hit Antwerp.[32] Together the V-weapons destroyed 200,000 buildings and killed nearly 10,000. Over 4,000,000 were made homeless. Hitler believed they would have been decisive if they had been operational in 1939.[33] The randomness of their targetry terrorized the populace. The Scuds of

Below : *The Me-410 was one of the aircraft used as heavily armed "bomber-shooters" against massed B-17/B-24 formations. These bomber-shooters were shot down in droves by USAAC P-51 & P-47 fighter escorts.*

Iraq engendered civilians with the same hopeless fear and defenses had improved only a little over the days of the V-2 (the "Patriots" were much less effective than implied by the press at the time). This technological trend continues.

Even though the V weapons presaged things to come, they were actually counter-productive as they diverted key war materials and used up precious skilled manpower that probably could have been better used elsewhere.[34]

Eastern Europe

From the Air War standpoint, nothing of real significance occurred in the east. 1944-45 just brought more of the same. More and more Russian aircraft of improving capability smothered the Luftwaffe in the east as the Germans pulled units back to fight the Battle of Germany with the Westen Allies. When the Russians got into Eastern Germany, and saw the devastation in Germany wrought by our heavy bombers, they became much more interested in strategic air power. We will see this realization in future chapters.

The Final Air War in the Pacific

The war in the Pacific, like that in Europe was really over in 1943, but the distances to Japan were just longer. Practically all of the major operations toward Japan were conducted in 1944. American domination was even more pronounced than in Europe. The sea was full of ships, mostly American, and the sky was full of airplanes, mostly American. In spite of grievous troop losses, the

issue was never in doubt and the capabilities of fighter air power drove the train. The air superiority of the F6Fs, F4Us, P-38s, P-47s, P-51s, etc., provided an umbrella for the amphibious attacks of the U.S. Army and U.S. Marines, as well as the attacks of USAAF and USN bombers.

Invasions of island complexes continued one after another. The Americans would choose an invasion site within land-based fighter range, establish air superiority, then using air superiority, invade; then build more airfields and leapfrog ahead within fighter range. Massive carrier raids assisted the land-based air to gain air superiority. Once air superiority was attained, the whole process would be repeated. The U.S. always tried to choose the least defended areas that would cut off Japanese bases and let them "die on the vine". MacArthur was particularly effective using this method along the New Guinea coast and up the island chains back to the Philippines.[35]

There were two major fleet engagements during 1944: The Battle of the Philippine Sea in June (the "Marianas Turkeyshoot")[36] and the Battle of Leyte Gulf.[37] Both were major U.S. victories. The loss of experienced Japanese carrier pilots was apparent in both. The sacrifice of Admiral Ozawa's carriers at Leyte Gulf showed Japanese awareness of the incapabilty of their pilots. Faced with the problem of no experienced combat pilots, the Japanese came up with a novel and deadly solution, the Kamikaze.[38] Suicide seems to fit the Japanese psyche and the Kamikaze was deadly to many U.S. ships. The first preci-

sion-guided bombs were a reality, but from the western viewpoint no one would call them "smart".

With the capture of Iwo Jima, P-51s could range the Japanese Islands and escort B-29s over their targets in Japan. U.S. Navy aircraft raided the home islands at will. The Atomic bomb then changed the world forever and its impact on war is still to be defined. As nuclear weapons were refined and made smaller, the fighter would become a viable delivery vehicle. World War II was over. The fighter had determined the victor and would again soon, but they would be powered by turbines engines and jet exhausts not propellers.

Summary

World War II was one of the cataclysmic events in all of history. It was the first war in history whose outcome was largely determined by air power. If air power was a strategic determinate of victory in WW II, then a key factor was air superiority as supplied by the fighters of either side. It is obvious that air supremacy was a determinant of victory in the end.

Tactically, the wisdom of WW I was perpetuated. Boelke's Dicta was as sound in WW II as in WW I when adjusted for context (the speeds and distances of the new aircraft and weapons). As we shall see, with another contextual adjustment this applies to jet age tactics as well. The fighting in pairs, line abreast formation with cross- over turns was as good in WW II as WW I, once one adjusted for the increased radius of turn dictated by faster, heavier fighters. Speed and altitude were as valuable as ever for energy to maneuver, and the age-old argument between speed and maneuverability still came out on the side of speed. Finally, the fighter pilot who first detects the other is usually still the victor.

When escorting bombers, it is better to use offensive fighters to disrupt defensive fighter concentrations before they can attack your bombers. Flying close formation with bombers makes your fighters nearly as vulnerable as the bombers. In most cases, slowing your fighters down to escort bombers just dissipates valuable energy, making your fighters vulnerable to enemy attack.

It was driven home that range is a key element of air power, particularly for fighters. Fighters could not achieve air superiority unless they could reach the targets and stay long enough to do their job.

In several cases, it was proven that back-up fighter and fighter systems should also be developed (the F4F saved the Navy when the Buffalo failed; the Mustang was originally a tertiary choice). Today, based on economic criteria, single production systems are often followed (the Pratt-Whitney TF-100 powered both the F-15 and F-16). The same type of decisions doomed Germany in WW II (Me-109 was the only operational German fighter during the Battle Of Britain). Have we drawn the correct conclusions from WW II?

Technologically, the progress made was staggering. We saw the perfection of the jet aircraft, radar, radio communication, and nuclear weapons. The Germans built not only viable jet fighters and bombers, but a rocket interceptor. Aircraft were able to detect other aircraft by radar and intercept them at night and in any weather. Aircraft had reached the stratosphere and could operate efficiently in near-space. Aerospace weapons had become a reality. The V-1 was the first true cruise missile with an indifferent guidance system and was accurate only as an area weapon. The V-2 was a terror weapon for which there was no defense (there was still only a limited defense in the Persian Gulf in 1990 where the "Scud" was nothing more than an improved V-2) still with relatively poor guidance systems.

During WW II, aircraft doubled in speed and range. Bombers, airliners and airlift ranged globally. Gunsights were designed to automatically track aircraft and solve most of the gunnery problem for the fighter pilot. Instrument landing systems were developed using navigation aids and radar. It was a technological explosion.

Above: *The Mig-3 was one of the myriad of fighters produced by the USSR after moving their factories east of the Urals. These aircraft were effective particularily at low-altitude. This production achievement was incredible.*

NOTES

1. Green & Swanborough, *The Complete Book of Fighters*, Smithmark Publishers, New York, 1994, p. 560.
2. Op. Cit.; p. 586.
3. Op. Cit.; p. 445.
4. Op. Cit.; p. 586.
5. Op. Cit.; pp. 215-216.
6. Op. Cit.; p. 318.
7. Op. Cit.; p. 409.
8. Op. Cit.; p. 327.
9. Op. Cit.; p. 164.
10. Op. Cit.; p. 456.
11. Op. Cit.; p. 299.
12. Boyne, Op. Cit.; p. 330.
13. Bekker, Op.Cit.; p. 344.
14. Ibid.
15. Op. Cit.; p. 350.
16. Bavouset, Glenn B., *More WW II Aircraft In Combat*, Arco Publishing, New York, 1981, p. 117
17. Op. Cit.; p. 353.
18. Stephen L. MacFarland and Wesley Phillips Newton, *To Command The Sky*, Smithsonian Institution Press, Washington, 1991, p. 162.
19. Adolf Galland, *The First And The Last*, Henry Holt and Co., New York, 1954, p. 187.
20. Raymond Tolliver & Trevor Constable, *Fighter General The Life Of Adolf Galland*, AmPress Publishing, Zephyr Cove, Nevada, 1990, p. 242.
21. Op. Cit., p. 243.
22. Boyne, Op. Cit.; pp. 336-338.
23. Op. Cit.; p. 336.
24. Op. Cit.; p. 249.
25. Bekker, Op. Cit.; p. 354.
26. Op. Cit.; p. 355.
27. Boyne, Op. Cit.; p. 349.
28. Op. Cit.; pp. 349-350.
29. Op. Cit.; pp. 350-351.
30. Ibid.
31. Ibid.
32. Ibid.
33. Eugene M. Emme, *The Impact Of Air Power: "Guided Missiles Could Have Won"*, D. Van Rostand Company, New York, 1959, pp. 260-267.
34. Boyne, Ibid.
35. Griess, Op.Cit.; pp. 175-184.
36. Thomas E. Griess, *The Second World War: Asia And The Pacific*, West Point Military History Series, Avery Press, Wayne, NJ., 1989, pp. 161-167.
37. Op. Cit.; pp. 189-195.
38. Op. Cit.; pp 198-199.

Above: *The Korean War ushered in the Jet Age. The Grumman F9F Panther was one of the effective jets that the U.S. Navy operated from carriers in Korea.*

The Lavochkin-LA-11 was a primary North Korean fighter at the start of the war.

CHAPTER 6

THE KOREAN WAR

The Initial Fight

When the North Korean Peoples Army (NKPA) attacked across the 38th Parallel on June 25, 1950, the world was completely surprised. Evidently the Russians were too. When the U.S. brought the aggression, to the U.N. Security Council, the Russians had previously walked out protesting some other U.N. decision. With the Soviets gone, the U.N. Security Council voted unanimously to support the U.S. in stopping the North Koreans as did the General Assembly. For the Communists, Korea was a mistake. The aggression by the Communists had unified the whole world for once.[1]

Following WW II, the U.S. Air Force had disarmed frenetically. Hundreds of fighters and bombers already in the Far East were literally chopped to pieces just a few months before they were needed again in Korea. The Army was in worst shape. There were four untrained, understrength, out-of-shape divisions in Japan and they were all that was available to stop 11 fully equipped and trained North Korean divisions pouring across the 38th parallel.[2] One unready division was all that was immediately available to deploy to Korea from Japan. That single division was thrown into the gap to delay the NKPA. That "puny" response was made by a nation that was arguably the most powerful in the world five years before.

The U.S. 24th Infantry division was thrown in front of the NKPA juggernaut piecemeal, companies, battalions, and regiments at a time. The division was deployed all over Japan. There was a shortage of shipping and even though the entire division was ordered to Korea there was no way to get it there. These troops were short of everything.[3] Task Force Smith, one of the first units deployed, had two understrength infantry companies, six 2.36 inch bazookas, two 75 MM recoilless rifles, and a battery of 105 MM with only six rounds of anti-tank ammunition per gun for its artillery.[4] Its experience was typical of all of the 24th Division units.

In its first combat, deployed in a driving rain, Task Force Smith learned immediately that the units' anti-tank weapons were worthless. The leading NKPA T-34 tanks were hit repeatedly with 75 MM recoilless fire and WW II 2.36 inch bazooka rounds with no visible effects even at close range. The tanks continued on into the artillery positions where three tanks were knocked out before the anti-tank rounds ran out. Air support could have knocked the tanks

out, but the heavy rain clouds eliminated that possibility. Task Force Smith delayed the NKPA advance for seven hours.[5]

The Air-To-Ground War

The only forces in position to immediately resist the invasion was the U.S. Far East Air Force (FEAF) commanded by Lt. General George Stratemeyer. FEAF was ordered to immediately attack the NKPA advance and did so. FEAF, unlike the massive force that destroyed Japan during WW II, consisted of only 1,172 aircraft of all types, eight combat wings; five fighter, two bomber, and one transport. But unlike the infantry they proved to be combat-ready.[6]

These forces were divided between offense and defense. Ten fighter squadrons were committed to the air defense of Japan, Okinawa, and the Philippines. This left the two bomb wings and eight fighter squadrons committed to Korea.[7] Within 30 days, the U.S. Navy Carrier Boxer delivered 145 F-51s to Fifth Air Force in Korea as reinforcements.[8] The U.S. Navy in the Pacific was also ordered to Korean waters. Immediately, on orders from the

LEFT: *The Mig-15 was a nasty surprise to U.N. Forces when China entered the war in late 1950. It outperformed all U.N. fighters until the USAF F-86 arrived in December 1950.*

97

Above: *The P/F-82 was two P-51's combined and in service in Japan when Korea erupted. Fast, it had very good range.*

President, FEAF began airlift to evacuate key personnel from Korea to Japan. Over 850 personnel were flown out with no casualties. The primary reason for this was the cover provided by Fifth Air Force fighters.[9]

At the same time, FEAF began striking NKPA forces south of the 38th Parallel and flying close air support missions supporting both U.S. and ROK forces. On 29 June, FEAF received approval to bomb north of the 38th parallel. By mid-to-late July, FEAF had established air supremacy. The North Korean Air Force (NKAF) was reduced to a shell of less than a score of aircraft.[10] Fifth Air Force had gained air superiority on day one and never gave it up except when the Chinese committed Mig-15s over North Korea, in November 1950 later in the war. When FEAF received F-86s, that temporary respite disappeared too (note that fighter aircraft again determined air superiority).

The close support for the U.N. Command (UNC) was highly effective and did severe damage to the NKPA forces on the ground. The air forces in Korea were commanded by Major General Earl Partridge, Commander of the Fifth U.S. Air Force in Japan. North Korean ground forces were attacked continuously wherever they were found as they moved to the south. Initially, as air-lifted troops of the U.S. 24th Infantry Division had no heavy weapons, close air support was also supplied to make up for this deficiency.[11]

By mid-July, a Joint Operations Center (JOC) to control close support was organized. Eighteen Tactical Area Control Parties were organized and did splendid service controlling strikes and close support against the advancing NKPA. A little later, light aircraft (North American T-6s) were introduced as "Mosquito" aircraft to control air strikes, both interdiction and close support (there had been some strikes on friendly forces in the confusion of the retreat).[12]

As the U.N. defenses began to solidify, more and more interdiction missions were flown. These attacks were highly effective against the advancing NKPA forces. For example, on July 10, 1950, 5th AF caught an NKPA column backed up behind a bridge which had been destroyed. They destroyed 117 trucks, 38 tanks, and 7 half-tracks.[13] The sole NKPA armored division arrived at the Naktong River in remnants, battered and decimated.[14] According to FEAF records, between July and October 1950, the FEAF killed 39,000 troops, destroyed 452 tanks, 75 bridges, 1,300 rail cars, and 260 locomotives (KIA figures were based on observed bodies so this count may be low).[15] Most of these losses were inflicted by FEAF fighter-bombers, and light bombers. At the same time all of North Korea's strategic targets were destroyed by heavy bombers (B-29s).[16]

According to General Weyland,

the commander of FEAF, interdiction had reduced the supplies needed by NKPA forces on the Pusan Perimeter to about one tenth of its needed daily tonnage (21.5 tons delivered daily vs. 206 tons required). He attributed the almost immediate collapse of the NKPA when the UNC broke out of the perimeter and the ease of the invasion at Inchon, to the NKPA's poor supply situation. FEAF attacks were the main reason for the NKPA's lack of supplies.[17]

When the Chinese Communist Forces entered the war on 6 November 1950, the FEAF was stronger than in July. FEAF by this time consisted of three B-29 wings, two B-26 wings, and 13 fighter squadrons. The fighter squadrons were made up of F-51s and F-80s.[18] The massive Chinese forces were

exceedingly vulnerable to air attack and suffered very high materiel losses and heavy troop casualties to FEAF action. Between November 1950 and June 1951, the Chinese losses to air were significant: 117,000 killed, 296 tanks, 80,000 shelters/buildings, 13,000 vehicles, 2,600 railway cars, and 250 locomotives.[19]

From July 1951 to June 1953, air was responsible for the destruction of more than:

- Vehicles 75,000

- Locomotives.....................1,000

- Rail Cars.........................16,000

- Bridges............................. 2,000

- Rail cuts.......................... 27,000
- Boats/Barges 600

- Troops 28,000

- Tanks 300

- Gun Posts.....................12,000

- Bunkers ....15,000 [20]

In the spring of 1951, General Ridgeway told President Truman he could go back to the Yalu for 100,000 U.N. casualties. Truman decided to accept an offer to negotiate by the Chinese rather than accept 100,000 more casualties. Over the next two years the U.N. would suffer losses of over 150,000 casualties and the U.N. would be stalemated in a repeat of WW I trench warfare in the Korean mountains.[21]

The Air War

On November 1, 1950, the Chinese introduced a swept-wing, high tailed fighter, the Mig-15, into northern Korea. Instantaneously, the air war turned completely around. The Chinese were now in the driver's seat. The Mig was faster than the F-80, could outclimb it, and almost turn with it. The B-29 was "meat on the table" for these Mig fighters. On November 8, General Vandenberg offered General Stratemeyer a wing of F-84Es and one of F-86As. The offer was quickly accepted by wire. Two wings of brand-new fighters were very much appreciated.[00] Again it would be useful to compare

Left: *The Republic F-84 Thunderjet was introduced in Korea in late 1950. The "Hogs" performed as air-to-ground aircraft from the time Sabre jets took over the air-to-air role. An F-84E is shown here. F-84s like all the straight-wing U.N. fighters were not equal to the F-86.*

opposing fighters:

- Chinese – Mig-15 (Fagot), Klimov RD-45 (copy of the Rolls-Royce Nene) 5,005# static thrust, 652 MPH, 1 x 37 MM cannon, 2 x 23 MM cannon. [23]

- USAF – F-86A (Sabre), GE J47-13, 5,200# static thrust, 601 MPH, 6 x .50 caliber machine guns.[24]

- USN – F9F-5 (Panther), Pratt-Whitney J-48, 6,250# static thrust, 604 MPH, 4 x 20 MM cannon.[25]

- Royal Navy – Hawker Sea Fury, Bristol Centarus, 2,400 HP, 460 MPH; 4 x 20 MM.[26]

It is interesting that present-day statistics list the Mig-15 as much faster than the Sabre. Pilot experience indicated the F-86 as slower than the Mig above 30,000 feet and faster at lower altitudes. The high tail (mid-rudder mounting) of the Mig might have limited the high angle attack capability of the plane. If the wing washed out the horizontal tail, the Mig-15 would enter "deep-stall", or "pitch-up". These terms define the uncontrolled pitch up of the nose due to loss of negative lift as the horizontal tail is blanked. I talked to an aeronautical engineer from ACD at Wright-Patterson, touting energy maneuverability in the 1960s in Europe, and he confirmed this hypothesis. I have never seen it confirmed or written in any document (the McDonnell F-101, Lockheed F-104, and Republic F-84F all had this problem to some degree). To pitch-up meant loss of control, and could mean bailout if at low altitude as it could take thousands of feet of altitude to recover.

In swept wing aircraft, this tendency is exacerbated.[27]

The development of both the F-86 and the Mig-15 are interesting as both aircraft benefitted from swept-wing technology researched and developed in Nazi Germany. For example, the F-86A was originally designed as a straight-wing aircraft. The straight wing version had difficulty meeting its 600 MPH specification. North American, the company that built the F-86, was aware of the German research on swept-wings which had been brought to the U.S. in 1945. The North American engineers were familiar with the high speed effects of swept-wings, but were puzzled about how to achieve low landing speeds with this type of design. At their request, a captured Messerschmitt wing assembly was shipped to North American. Messerschmitt 262 used leading edge slots to reduce landing speeds. North American incorporated this type wing assembly and the results were spectacular. The F-86 became a true trans-sonic design that could exceed the speed of sound in a dive.[28]

The F-86 used six .50 caliber machine guns. According to author Richard Hallion, four 20 MM guns were tried and the gases from their muzzle blast was ingested into the Sabre's GE engine and induced compressor stalls, so the decision was made to continue with the .50 calibers.[29]

By December of 1950 the F-86s from the 4th Fighter Wing were in com-

bat. On its first combat mission Colonel Bruce Hinton destroyed a Mig. The swept-wing jet war for the air over Korea was on.[30] Hinton was lucky. Because of the distance from Kimpo to the Yalu, the Americans were trying to conserve fuel and were patrolling at .62 mach. When he spied the Migs, they were below the F-86s and the Migs took no defensive action (probably assuming the "bogies" were F-80s which were not a threat with a .76 mach top speed). Hinton dove on the Migs and made up his low mach before he engaged. It took about three days for the Sabres to realize that they had to patrol at .85 to .87 mach. There was no time to accelerate in a fight particularly when they were being bounced from above (it was soon apparent that the Mig-15 had a higher ceiling than the heavier F-86). It was imperative that the energy already be there when the fight started.[31]

By the end of December, the 4th Wing could take pride in its achieve-

ments. The wing had flown 234 sorties and 76 Sabres engaged Migs, destroyed eight, probably destroyed two, and damaged seven others. The Wing had developed optimum tactics which varied only slightly for the rest of the war. 4th Wing used a "jet stream" method where Sabre flights arrived at five minute intervals which provided a minimum of four, four-ship flights always within supporting distance of one another. These were basically Rotte and Schwarm tactics varied to adjust to the wider turns required by the higher speeds of these trans-sonic jets. As indicated above, the flights maintained high mach speeds which allowed the F-86s to maneuver offensively against any attack or opportunity.[32] The Sabres had restored air superiority over Northwestern Korea.[33]

In January 1951, when Kimpo and Suwon Airfields were threatened as the CCF (Army) marched further south, Fifth Air Force withdrew its Sabres back to Japan. With no F-86s to oppose them, the Migs became increasingly active over North Korea attacking F-80 and F-84 fighter bombers. Over Sinuiju, 33 F-84s were attacked by 30 Mig-15s. The Migs showed superior speed, climb, and acceleration, but the Thunderjets shot down four Migs. In addition, the Thunderjets strafed Sinuiju and Pyongyang was bombed by F-80s and B-29s. No bombers or fighter-bombers were lost.[34]

UN air superiority was continually reduced during January and February 1951 with the Sabres in the barn in Japan. Why the CCF did not follow up with aggressive attacks on our fighter bombers is still a mystery today. It was during this period that the area between the Chongchon and Yalu rivers was dubbed "Mig Alley".[35]

The last four RF-80 reconnaissance flights in February were attacked by Migs and barely got home. The single Fifth Air Force kill in February 1951 was a Yak-9 shot down by an F-51 Mustang on 5 February. When the Eighth Army drove the Chinese back north of the Han, General Partridge began to stage through Taegu and Suwon even though both bases were in very poor condition. As the Sabres again appeared over Mig Alley on 1 March, the American pilots found that the Migs were now very aggressive, but still poor shots.[36] With only two squadrons (max of 50 aircraft), the Americans were outnumbered. Constantly reinforced, the CCF had 445 Migs in Manchuria by June while the Sabres had no more than 89.[37]

In June, the Migs became very aggressive. What the U.N. pilots called the "Honcho" pilots were flying the Migs (Honcho means boss in Japanese). Wingmen covered their leaders and the Migs changed tactics that used the Mig's strengths. Instead of diving through at high speed as they had in the past, they now pulled up and climbed back to high altitude.[38]

By September there were over 500 Migs in Manchuria and the Communists again made a bid for air superiority. Using the procedures developed in June the Mig pilots attempted to exploit the strengths of their aircraft above 35,000 feet. The Sabre pilots were outnumbered three or four to one, but still attacked when they could. The F-80s were still shooting down Migs at a ratio of about 10-1, but Migs were getting through and disrupting the interdiction plans by the U.N. fighter bombers. Fifth Air Force was under no illusions and asked USAAF Headquarters for more F-86s. The reply was none were available, nor was there any support available for another F-86 Wing in the Far East.[39]

October 1951 was grim indeed for Fifth Air Force. The 4th Wing knowing it was the only game in town aggressively attacked the numbers of Migs wherever they found them. They were outnumbered in every battle, yet shot down 19 Migs by 16 October.[40]

About this time three new airfields were discovered under construction in North Korea. If these airfields were completed the Mig radius of action would extend down the peninsula at least to Pyongyang. The fighter bombers interdicting railroads and lines of communication were often interrupted by Migs as the large numbers slipped by the F-86s. The F-80s and F-84s gave as good as they got in these forays but the Mig was so superior that some fighter-bomber losses were a given.

The prop bombers (B-29s and B-26s) were decimated if caught without F-86 support. Night attempts to neutralize the new North Korean airfields were ineffectual, so the B-

29s had to attack them in daylight in the face of the overwhelming Mig presence. Air superiority in Northern Korea was clearly in jeopardy. For example, on October 23, 24 B-29s escorted by 55 F-84s went after the airfields. This attack was screened by 34 Sabres over Mig Alley. The Sabres were attacked and boxed in by 100 Migs. Two Migs were shot down, but the Sabres were out of the bomber fight to the south where 50 Migs avoided the slower Thunderjets and pounced on the B-29s. Three Superforts went down and 20 others were damaged. Only one bomber escaped damage. Every raid was roughly handled and the airfields were still operational.[41]

During October 1951, the Communists lost 24 Migs to Sabres, 7 to B-29 gunners, and 1 to an F-84. But U.N. losses were 7 Sabres, 5 Superforts, 2 Thunderjets, and one RF-80. The B-29 force had lost 55 aircrew killed or missing, 12 men wounded, and eight B-29s had been heavily damaged. Migs and conventional communist aircraft were in revetments in North Korean airfields south of the Yalu for the first time.[42]

To complicate the situation, in December, the Reds introduced the Mig-15bis (bis means encore). This new version was powered by a new VK-1 engine of 6,000 #s of static thrust, was more stable, and had an even higher altitude capability.[43] In the meanwhile, the USAF was upgrading the F-86A to the F-86E. The only difference between the "A" and "E" was the all-flying tail (the horizontal stabilizer and elevator of the A were converted to a stabilator where the whole horizontal tail moved. This provided the "E" with much more pitch authority trans-sonically and was a feature of all later sonic aircraft).

HQ USAF provided 75 more F-86s from Air Defense Command and FEAF converted the two squadron 51st Wing from F-80s to F-86s. After a very short transition period, the 51st flew its first combat missions on 1 December 1951. FEAF now had 165 Sabres.[44]

In the meanwhile, the B-29s had turned to night bombing using Shoran, MSQ, and radar to assist them in hitting the targets. The new airfields in North Korea were

Below: *The T-33 was a two seater development of the F-80 Shooting Star. It was the basic trainer of the USAF for years. Lockheed made thousands of these trainers which were used by many Air Forces worldwide.*

102

reduced. U.N. fighter-bombers were occasionally attacked but air superiority returned to the Sabres as they attacked every Mig that moved and the Communists pulled back after March. They sustained numerous losses.[45]

The Navy Air War

The USN aircraft did not participate much in the air-to-air combat in Mig Alley. Its best jet fighter, the Grumman F9F-5 Panther, turned out to be slower, less maneuverable, and could not climb or dive with the Mig.[46] Nevertheless, a Panther pilot outmaneuvered an attacking Mig-15 and promptly shot him down. The four 20 MM guns were deadly if the Panther got into firing position.[47] F4U Corsairs flying interdiction missions were frequently bounced by Migs and if the Migs chose to maneuver with the Corsairs, they found the F4Us had 20 MM teeth. Several Mig-kills were made by F4U drivers who stayed low and forced the faster Mig to try to turn with the more maneuverable Corsairs.[48] Unfortunately, all the Mig pilots were not that stupid and like USAF F-51s, the F4Us were meat on the table to experienced Mig pilots. Luckily, these experienced Mig pilots were

few and far between and the UN prop jobs usually got away.

The biggest air-to-air contribution was made by USN/Marine aircraft at night where F7F and F4U night fighters were used to intercept the PO-2 "Bedcheck Charlie" harassing flights by the North Koreans. This maneuverable old biplane was difficult to intercept often flying at less than 100 MPH. Many of these interceptions were successful however, using these Navy prop-driven night fighters.[49]

Less well-known was the night work of the twin engined Douglas F3Ds who flew both night interdiction and night fighter missions. When CCF night fighters began to attack USAF B-29s, F3Ds were employed to escort the "Big Friends" at night and attacked these communist night fighters - again with some success. During November and December, this night war raged. In spite of USAF F-94s joining the fray, and numerous Communist night fighter losses, a number of B-29s were lost to the enemy night fighters. During January 1953 however, the corner was finally turned in some desperate fighting. From the end of January until the war

ended no further bombers were lost.[50]

In June of 1952 the USAF introduced the F-86F. This aircraft differed from the F-86E as it had a larger engine of 5,900 #s of static thrust and a hard leading edge on the wing. It was 10 knots faster than the "E", and could outturn it. In addition, it had a 53,000 foot ceiling. American pilots thought the F-86F the equal of the Mig-15/Mig-15bis. The F-86F was first introduced in two squadrons, but as its superiority was noted by all the American pilots, the aircraft was parcelled out to all of the squadrons.[51] Let us consider the following lineup as of June 1952:

• USAF – F-86F Sabrejet, J47-GE-27, 5910 #s static thrust, 678 MPH, 6 x .50 caliber machine guns [52]

• CCF – Mig-15bis, VK-1, 5,952 #s static thrust, 688 MPH, one 37 MM cannon, 2 x 23 MM cannon.[53]

The hardwing on the F-86F was a closely guarded secret in 1952. Anxious to test and analyze a flyable Mig-15, the USAF offered $100,000 dollars for a Mig-15 in flyable condition. American pilots promptly dubbed this "Project Moolah".[54] During this time a package of four 20 MM cannons was tried on eight F-86s. The package was found suitable but needed refinement before it was adopted in

combat (probably because of the ingestion problem mentioned by Hallion).[55] The 20 MM cannon were eventually incorporated into late model F-86Hs. The F-100 was the first USAF fighter designed from scratch with 20 MM cannon. The spring of 1953 was feast time for the Sabre pilots. With a better and faster mount the younger, aggressive pilots cleaned up on any Migs that would fight. In May the F-86s sighted 1,507 Migs, engaged 537, and shot down 56 for the loss of only one Sabre. Increasingly, Mig pilots were seen to lose control of their aircraft and bailout during hard maneuvering.[56]

For some reason in June of 1953 the Migs descended below 40,000 feet where the Sabres were at their best and 77 Migs

were destroyed for no USAF losses. In July the Sabres got 32 Migs for four losses.

Summary

The fledgling USAF came of age in Korea. Grabbing air superiority on July 29,1950 the USAF/UN never really relinquished it although it was touch and go during 1951 on more than one occasion. The USAF that emerged from the war bore no resemblance to the small force that began it.

The overall kill ratio of the F-86s over the Mig-15 was 14-1 in favor of the F-86. Air force fighter doctrine was codified for the first time in double ace Major Fred "Boot" Blesse's "No Guts, No,Glory" (see the attached appendix). Boelke's pairs and line abreast tactics survived the tests of jet warfare. Capture of a Mig-15bis after the war revealed that the Mig was not as good as we had thought when fighting them. It had high speed control problems, poor high speed stall

characteristics (pitch-up?), inadequate defrosters that limited visibility, and poor lateral and directional control at high altitude to name just a few.[57] Anytime a kill ratio of 14-1 is run up, you can bet the top dog has a better airplane. Pilots as a group simply are not that much better than another group of trained pilots. Pilot survivors in combat usually equalize quickly. Cherchez the best fighter (weapons system) where the scores are highest.

This F-86's superiority preserved overall air superiority which allowed the UN air forces to provide the ground forces with superior close support and interdiction throughout the war. Note that the Mig-15 and F-86's range played heavily in the strategy of the war. If the Mig-15 had more range the war might have been very different in 1951. Fighter air refueling became a reality well after Korea in the mid to late 1950s. A new dimension was thereby added. It began in Britain by the way – and has been a factor in fighter operations ever since.

NOTES

1. T.R. Fehrenbach, *This Kind Of War*, The MacMillan Company, New York, 1964, p. 702.
2. Op. Cit.; pp. 92-94.
3. Op. Cit.; p. 98.
4. Ibid.
5. Op. Cit.; pp. 100-108.
6. James T. Stewart, *Airpower, The Decisive Force In Korea*, D. Van Rostrand Company, Inc., Princeton, N.J., 1957, p. 6.
7. Ibid.
8. Op. Cit.; p. 7.
9. Frank Robert Futrell, *The United States Air Force In Korea 1950-1953*, Van Rees Press, New York, 1961, pp. 12-13.
10. Ibid.
11. Stewart; p. 9.
12. Op. Cit.; p. 109.
13. Op. Cit.; p. 9.
14. Ibid.
15. Op. Cit.; p. 10.
16. Op. Cit.; p. 81.
17. Op. Cit.; p. 9.
18. Op. Cit.; p. 12.
19. Op. Cit,; p. 15.
20. Op. Cit.; table; p. 25.
21. Mathew B. Ridgeway, *The Korean War*, Doubleday & Company, New York, 1967, pp. 150-151.
22. Futrell; Op. Cit.; p. 232.
23. Green and Swanborough, *The Complete Book Of Fighters*, Smithmark Publishers, New York, 1994, pp. 390-391.
24. Op. Cit.; p. 449.
25. Op. Cit.; pp. 266-268.
26. Op. Cit.; pp. 289-290.
27. H.H. Hurt, *Aerodynamics For Aviation Personnel*, Aerospace Safety Division, University of Southern California, Los Angeles, 1964, p. 313.
28. Futrell, pp. 233-234.
29. Richard P. Hallion, *The Naval Air War In Korea*, Kensington Publishing Corp., New York, 1988, p. 229.
30. Futrell, Op.Cit.; pp. 234-235.
31. Ibid.
32. Futrell, Op. Cit.; p. 236.
33. Op. Cit.; p. 237.
34. Op. Cit.; pp. 268-269.
35. Op. Cit.; p. 269.
36. Futrell, OP. Cit., pp. 270-270.
37. Op. Cit.; p. 372.
38. OP. Cit.; p. 281.
39. Op. Cit.; pp. 370-399.
40. Ibid.
41. Ibid.
42. Ibid.
43. Op. Cit.; p. 380.
44. Ibid.
45. Ibid.
46. Hallion, Op. Cit.; p. 246.
47. Op. Cit.; p. 247.
48. Op. Cit.; p. 241.
49. Op. Cit.; pp. 254-258.
50. Op. Cit; pp. 269-277.
51. Futrell, Op. Cit.; p. 609.
52. Grenn & Swanborough, Op. Cit.; p. 452.
53. Op. Cit.; p. 392.
54. Op.Cit.; p. 610.
55. Op. Cit.; p. 609.
56. Op. Cit.; p. 612.
57. Op. Cit.; pp. 650-651.

Top: *The F-101B was a premier intercepter during 1950's–60's. It was one of the all-weather fighters designed to intercept nuclear bombers. It was armed only with missiles with no gun. (Falcon missiles & Genie nuclear missiles).*

Bottom: *The USSR like the USAF continued to develope two-seaters of primary fighters to train new pilots. The two-seater version of the Mig-21 is shown here, Nato codename: Mongrol.*

CHAPTER 7

THE FIGHTER IN THE VIETNAM WAR

INTRODUCTION

In the 12 years between the Korean and Vietnam Air Wars (1953-1965), there were several monumental developments in fighter aviation:

In the USAF, the "Century Series" aircraft were introduced; the F-100, F-101, F-102, F-104, F-105, and F-106 followed one another in quick succession, and all were supersonic in level flight.[1] This was the most extensive period of peacetime fighter development in U.S. history. The US Navy contributed with its first two supersonic entries: the F-8 Crusader and the F-11 Tiger.[2]

In France and Britain, supersonic fighters similar to those in the U.S. were also being developed. Dassault in France developed a series of delta-winged fighters that are still with us today through continued development. The "Mirages". Perhaps the most famous of this series were the Mirage IIIs used by Israel against the Arabs in the 1967.[3]

In Britain, the RAF developed the English Electric "Lightning" a distinctive mach 2 interceptor with two afterburning engines mounted one over the other similar to an over-and-under shotgun.[4]

Russia provided the opposition for all of these free world supersonic fighters with improved subsonic Mig-17s, supersonic Mig-19s, Mig-21s,[5] Su-7s, and Su-15s. Typically, the Soviet designs tried every wing platform on the same basic air-frame. During this period, they used both delta wings and low aspect swept wings on most of their designs. Then they adopted the design that worked best for production. Many were sent to third world countries. The three Migs saw combat in the Middle East, India, and Vietnam. The Su-7 fought in the Middle East and India. Su-15s were used in Soviet air defense exclusively and never saw fighter versus fighter combat.[6] Of the third world aircraft, the Mig-21 proved to be the most formidable in combat.[7]

The aerodynamics of all these aircraft were similar to each other, but differed markedly from most of the previous subsonic designs. We will analyze the aerodynamic differences using U.S. aircraft, but the analysis applies across the board.

Another revolutionary change, was the introduction of mid-air refueling. Conceived in the 1920s by the U.S. Army Air Corps, mid-air refueling was developed and perfected in England in the mid-1950s. The British system was adopted quickly by the USAF and by the late 1950s U.S. fighters were routinely flying across both oceans without landing. If tanker aircraft were available, range as a limiting factor in fighter operations was no more. World wide range had become a reality.[8]

Weapons systems also changed dramatically in this period, as we entered the air-to-air missile era for the first time. These missiles were guided by radar or homed on Infra-red (heat) radiation. Their initial introduction led to some mistakes in fighter armament as we shall see, but these weapons are still a major technologi-cal trend today and for the future.

The major change in fighter performance was the increases in engine performance provided by afterburning. Before we discuss Vietnam, it will be useful to analyze all of these changes.

AFTERBURNING/REHEAT

The major new characteristic of the century series aircraft was the use of afterburners (ABs). It was this engine feature that allowed these fighters to be supersonic vehicles. Turbo-jet engines operate with airflows far in excess of that necessary to support combustion. The excess air is used mainly for cooling engine turbines and hot section parts. As this excess air passes through and around the engine, it is heated to very high temperatures. By using the hot, unoxidized oxygen of this excess air in the tailpipe and injecting fuel directly into it, the turbo-jet engines of the Century series could gain 40-60% more power from their engines. This additional power was attained without having to increase the already critical temperatures in the hot sections of the main engine. The added combustion chambers and nozzle in the tailpipe increased engine weight only about 10-20%.[9]

None of the Century Series aircraft could negotiate the large drag rise encountered at Mach 1 in level flight without this additional afterburner thrust. Therefore, they needed afterburners to go supersonic in level flight. But this increase in

107

power was not without costs. In reheat, (British for afterburning), fuel consumption increased by factors of from 2-1 to 4-1.[10] As a worst case example, for an F-101, if the fuel consumption quadrupled with the 'burner' on, overall fuel consumption would have increased by a factor of eight as the F-101 had two engines. Supersonic flight demanded some big-time fuel flows.

With this huge fuel consumption in AB, pilots were forced to use afterburner carefully. Running out of fuel in any airplane is serious, but in these planes even more so. None of the Century Series aircraft were recommended for "deadstick landings" (landings with no power). If you ran out of gas, you bailed out.

The Century Series aircraft were very slow to accelerate from transonic to high supersonic speeds, and afterburner had to be used extensively to attain maximum Mach. By the time most reached their listed top speed, (e.g., Mach 2 for the F-105) they were out of gas and either had to air refuel or return to base. Even though they were technically supersonic airplanes, air combat was obviously largely sub-sonic with these aircraft. All of the Century Series flew in Vietnam. The F-102, F-104 and F-106 saw very little or no combat as they were primarily used for air defense in-country.

Afterburners also required engine nozzles to optimize exhaust flow. Once the afterburner ignited, to accommodate the increased exhaust flow through the tailpipe the nozzles on the end of the tailpipe had to open to a supersonic flow position. Most aircraft used fuel pressure or hydraulics to automatically adjust the nozzles when the afterburner was selected by the pilot. Failure of the nozzle to operate properly gravely affected engine performance. Failure of the nozzle system to open when the afterburner was selected could cause the engine to compressor stall (the engine flow reverses and blows back from the tailpipe to the turbine which stalls or stops the compressor). If the afterburner failed to light with the nozzle in the open position, it significantly reduced the thrust of the main engine (about 50% if the nozzle could not be reclosed). Both were serious: Compressor stalls could overheat and destroy the engine and reduced thrust at a critical time (e.g., on takeoff) could cause a crash (see the landing pattern discussion below).

Afterburners have one other drawback. As they are an enormous heat source, they provide a huge IR target for infra-red guided missiles and sensors. Fighter pilots in my day used to joke about defensive tactics when you were the Blue flight leader. When an IR missile was fired at the flight, the technique was to note the path of the missile, ensure it was launched at your flight, then to call "Blue Flight, Burners now!" As your wingmen went into burner, you would stopcock your engine! (shutting down the heat source from your engine so the missile would home on your wingmen in afterburner). You will find if you stay around fighter pilots long that they all love gallows humor.

SUPERSONIC AERODYNAMICS AND PROCEDURES

Following the penetration of the sound barrier by the Bell X-1 rocket, the unknowns of supersonic flight were explored and mastered. The original sub-sonic equations for lift and drag went to infinity when they reached the speed of sound as they assumed air to be incompressible.[11] All the equations had to be reworked

Right: *USSR designers corrected most of the deficiencies of the Mig-15 in the Mig-17 (shown here). Mig-17's proved formidable opponents to "supersonic" aircraft over N. Vietnam. Their subsonic maneuverability and high thrust to weight ratio were difficult to counter in Century Series aircraft who far from home could not use their supersonic capabilities without running out of fuel or dropping their ordinance.*

based on transonic and supersonic test data.

The North American F-100 Super Sabre was the first supersonic fighter in the USAF.* It broke ground for the century series aircraft to follow. One major problem area, which assumed major proportions immediately, was that the low aspect ratio, swept-back, supersonic wings dramatically changed landing techniques. Supersonic airfoils were symmetrical, thin, and developed lift in an entirely different way from their sub-sonic counterparts. This difference was particularly exacerbated in the landing pattern and at the low speeds necessary for landing. To develop lift, these airfoils had to fly at positive angles of attack and at landing speeds, the angle of attack needed to be quite high in order to develop enough lift to safely land. As these severe angles of attack neared the stall, areas of increasingly high drag were encountered.

When combined with swept wings for transonic flight, landing performance suffered even more as swept wings are inherently very high in drag at lower speeds. With this high drag and very high angle of attack on landing, the F-100 required an entirely new landing technique.[12] If the airspeed were allowed to get too low and the wing approached to the stall, the aircraft would realize a rapid increase in drag and a correspondingly high loss of lift which could be dangerous. This necessitated power-on patterns all the way to landing.

Traditional overhead approaches had to be made loosely, with a very flat, power-on final approach. In F-100As and Cs, which had no flaps, the landing attitude was so nose high it was easy to drag the tail on landing and landing speeds were relatively high (180-200 knots). A tailskid was installed to prevent damage to the afterburner nozzle on nose-high landings. Anti-skid brakes were installed to accommodate the higher landing speeds and a drag chute system installed to quickly slow the aircraft on landing roll without excessive use of the brakes. The addition of flaps on the F-100D and F-100F lowered the nose-high landing attitude and permitted landings at slower speeds down to about 135 knots, but the flaps increased drag even more, requiring even more power.[13]

"Bank and yank" F-86 pilots, who liked high "G" landing patterns got into immediate trouble with the F-100 ("Hun") Others who reduced the power to idle up on the final approach, had accident after accident when transitioning to the "Hun". With a steep final, or no-power final, it was very difficult to land the F-100. Without power, the aircraft would rotate, but in spite of a nose high attitude the Hun

*Note: A supersonic airplane is defined as one that can accelerate to supersonic flight in *level flight*.

109

would often continue to sink straight into the ground before power could be re-applied.[14] The approach needed to be flatter and the sink and drag rate controlled with power. Once the pilots were shown the differences, it was no sweat, but it took time to acclimate the F-86 drivers accustomed to the traditional flare to the "Supersonic Super Sabre".

To make matters even worst, the J-57 (F-100, F-102, F101, F-8) and J-75 (F-105, F106) engines required a number of seconds to "spool-up" from idle to 100% "military" power (max power w/o burner). The time to spool-up from 75-80% power to military power on the other hand, was very short. On a correctly flown century series approach 75-85% power was used all the way to touchdown.

Afterburners were not reliable enough to be used in the traffic pattern or on landing. If one was foolish and slammed the throttle forward and outboard to go directly into afterburner in order to recover a bad landing, one invited disaster:

First of all, the engine would compressor stall if the air coming into the intake at landing speeds was insufficient to maintain a positive airflow pressure from the compressor in the intake portion of the engine all the way to the afterburning exhaust. If the afterburner exhaust pressure got higher than

Below: *Late-model F-100Ds. Note the curved air refueling probes. On early model "Ds" this probe was straight and refueling required experience and some skill as the drogue would jump up and to the right due to wash off the wing & fuelage.*

that in the forward parts of the engine, the burning exhaust would blow back out into the front of the engine. This stops the engine momentarily and can easily overheat it as the cooling airflow is disrupted. If severe, compressor stalls might even destroy the engine. To the pilot, compressor stalls felt like engine explosions and sometimes would scare the hell out of you.

Another problem that could occur with rapid throttle movement into afterburner was that the afterburner might not light. As the nozzle in the F-100 was positioned by fuel pressure pumped to the afterburner, the nozzle normally opened a split-second before the burner lit. If for some reason the burner failed to light, and the nozzle remained open, military thrust was severely degraded. If this happened in landing configuration with the gear and flaps (if equipped with flaps) down, there was insufficient thrust to maintain flight.

The other problem with using the burner when landing is the nozzle might fail closed when the after burner is selected. This failure could cause a "hard light" as the afterburner would light and then blow the nozzle open to the AB position. (the pilot would feel a hard jolt when selecting burner). If the nozzle eyelids did not blow open and still remained closed after the afterburner lit, it could cause a massive compressor stall and engine explosion, or meltdown.

With any of these failures, there was insufficient thrust to fly. Consequently, we used only military power for "go-arounds" and "wave-offs" for the F-100. This meant in the F-100D that there was only about 11,700 pounds of thrust available to get out of trouble when landing, versus the max power with afterburner of 16,900 pounds.[15] This throttle limitation was a characteristic of all the century series aircraft except the F-104, F-11, and F-4. The J-79 engine on these aircraft had an automatic fuel control which allowed the pilot to slam the throttle into military power with no problem. The J-79 also had a graduated afterburner that cut in using incrementally smooth segments, but again the burners were not recommended for go-arounds or waveoffs.

The expression "behind the power curve" is often used today, even in business, to denote an unrecoverable position. The expression originated in the early days of Century Series aircraft where it was possible to have aircraft drag exceed available thrust by initially trying to land a Century Series aircraft from a steep approach without sufficient power.

The McDonnell F-101A was a two-engined development of the XF-88 Voodoo. Developed as a long-ranged escort fighter to accompany SAC bombers, the F-101A was not really successful as a fighter-bomber. McDonnell's design did not adapt well to the ground attack role. But the aircraft was a performer and proved very flexible. We eventually ended up with three versions: The F-101A fighter, the RF-101 reconnaissance version, and the F-101B interceptor.

It must be remembered that in the 1950s USAF tactical fighter aircraft were planned to deliver atomic weapons (F-100s, F-101s and F-105s all concentrated on atomic deliveries like a small SAC). The F-101As had long range, high speed and good carrying capacity, all perfect characteristics for a nuclear penetration and delivery vehicle.

These features were also required for a good reconnaissance aircraft. When a reconnaissance payload was installed, the '101 became an outstanding reconnaissance aircraft. In a reconnaissance role, the RF-101 proved invaluable, and it was operational for over 20 years from 1954 until 1975. RF-101s served very well in Vietnam until replaced by RF-4s beginning in the late 1960s.[16]

However, the F-101A fighter version proved as poor in the fighter-against-fighter role as it was as a fighter-bomber. The aircraft was severely hindered by its lack of maneuverability. With a small wing and a very high wing loading, it had a limited turn capability. The high T-tail made the aircraft susceptible to pitch-up at high angles of attack, a particularly bad characteristic for an air-to-air fighter. The low-mounted stabilators on the F-4 (the next successful McDonnell design) were no accident as this low installation avoided pitch-up problems.

However, with its high speed, rapid acceleration and outstanding rate of climb, the F-101 had all the characteristics of a great interceptor. As an interceptor for bombers, maneuverability was less important and with a few design changes the F-101s were unexcelled in the interceptor role. A good radar was installed and a radar intercept officer added to make the F-101B into a dedicated interceptor. Armed with a nuclear-capable Genie missile and three Falcon air-to-air missiles, the two-seater F-101B served well in this role.

The Convair F-102 had a Delta platform that avoided the landing difficulties with supersonic airfoils. As the delta sinks into the runway and enters ground effect very near the runway, the air under the wide triangular wing compresses to cushion it to a very soft landing. The F-102, however, had other problems. With the same engine as the F-100, its transonic drag was too high and the airplane could not get through the mach. (There is a very large drag rise at mach 1 – the so-called "sound barrier".) The design was saved by an aeronautical engineer at NACA (now NASA), named Richard Whitcomb. Whitcomb's "Area Rule" showed that transonic drag was reduced when the cross-sectional area was kept constant in a smooth curve. The easiest way to do this was to pinch the fuselage in where the wings attached to the fuselage. The "Deuce" went supersonic the first time the aircraft flew after the Area Rule was incorporated.[17] This wasp-waisted technique was also used on the Convair F-106 and Republic F-105.[18]

The F-104 was a different design entirely and reached Mach 2 while other fighters were flying at much lower speeds in 1954. Its small, straight razor-sharp wings were unique. It was called "a missile with a man in it". With its small wing, it was a streamlined "rock" if the engine quit. The GE J-79 engine had many problems early-on and the procedure for loss of power was ejection. Because of its high mounted horizontal tail initially a downward ejection seat was

Above: *The McDonnald-Douglas A-4 was one of the most successful "point" designed aircraft. "Point" designed aircraft have optimized performance at high subsonic mach. Designed specifically for this regime, the approach optimizes performance, payload and range in the .8 to .9 mach range.*

installed. Unfortunately, the downward ejection seat turned out to be a poor design and was replaced after a complete redesign with an upward ejection system. As mentioned in our earlier discussion of the Mig-15 and F-101, the high tail which gave low drag at cruise and unaccelerated flight, also had problems with "pitch-up" in accelerated turns or pull-outs. Below 15,000-20,000 feet the emergency procedure for pitch-up was bail-out. Like the F-101, the Starfighter had a stick-shaker and kicker to prevent pitch-up. Imagine being tracked by an enemy fighter below 20,000 feet and needing to tighten the turn with the shaker and kicker telling you if you tighten the turn you will pitch up and go out of control! Only 153 F-104s were bought by the USAF. Another 2,400 were sold overseas after most of these aircraft problems had been cured.[19]

The F-105 Thunderchief was designed as a nuclear penetration fighter which could carry a nuclear weapon internally in a bomb bay with a backup capability as a fighter bomber. The "Thud" was to bear the brunt of "Rolling Thunder" fighter-bomber operations in Vietnam.

The F-106 was one of the best of the bunch. It was a superb all-weather fighter, and was fast, maneuverable, and easier to fly than the other Century Series. However, it was designed to intercept bombers and like the F-102 and F-101B initially carried no gun. The missiles it carried were not effective in hard G, fighter-to-fighter fights. In the late 1960s, when the lack of a gun was realized as a serious deficiency, the F-106s were retrofitted with 20 MM Gatling guns.[20]

The F-104, F-105, and F-106 were Mach 2 aircraft at top speed, but their afterburners burned so much fuel reaching those speeds that the planes just about had to turn around and land after reaching them. The F-100, F-102, and F-106 were the only Century Series fighters with a real dogfight capability and only the F-100 carried a gun. That brings us to another important development between wars: air-to-air missiles.

As pointed out above, the fighter forces after the Korean War were divided into the all-weather interceptor school and the tactical fighter school. The interceptor people were dedicated to shooting down nuclear bombers some of which were supersonic. The tactical fighters were interested in penetration and delivery of small "nukes". In the nuclear age, fighter aircraft were supposedly too fast to permit dogfights just as Spitfires and Hurricanes were supposed to be too fast for dogfighting prior to WW II. Nuclear war was deemed to have made dogfighting irrelevant.

In the USN, the F-8 was one of the best supersonic dogfighters of the day and a really superior design. By drooping the leading edge of the wing, Vought achieved an efficient trans-sonic airfoil while preserving excellent slow speed characteristics. Vought also achieved outstanding landing characteristics by hydraulically cranking down the F-8's fuselage and nose for landing, while the wing remained at high angles of attack to produce lift. This was an innovative solution to the nose-high landing problems of the day's supersonic swept-wing machines. Imagine trying to land a F-100 as described above on a carrier. The Vought F-8 was an incredibly good carrier fighter design. The shoulder mounted high wing and low mounted stabilator also had many supersonic stability advantages. Armed with four 20 MM cannon, and two-four AIM-9 Sidewinders, the late model F-8s were mach 1.8 aircraft and could maneuver with the Migs in Vietnam. In addition, F-8 pilots were specialists in air to air combat. It was no accident that the F-8 had the highest kill ratio of any aircraft in Vietnam.[21]

It is interesting that the Navy also bought off on the missile Pk myths in 1958 when it chose the F-4B Phantom II over a follow-on to the Crusader. The F-4B had no gun. Missiles were supposedly so deadly, the gun was just dead weight.[22]

Meanwhile, USAF air to air training virtually disappeared. The only air combat maneuvering (ACM) we could get was illicit and unbriefed. One literally risked his wings to "bounce" (attack) another aircraft. I remember being so desperate that I bounced a USN F3H Demon (the Navy guys always liked a fight and kept their mouths shut) in the Philippines in an F-100 while carrying a 2000 pound simulated "Nuke" on my left intermediate wing station which restricted me to 4 positive "Gs". Happiness was flying a test hop when the ship was clean and I could really learn something. Air combat maneuvering (ACM) between aircraft of different types was banned in the USAF as it was considered too

Above: *The Convair 102 ("Duece") was our first supersonic intercepter, but it required the discovery of the "area rule" by Richard Whitcomb at NACA (NASA) to be successfuel.*

dangerous and might cause accidents. This was changed in the late stages of the Vietnam War, but it is incredible even today, to recall how hard it was to get this change made. There were many losses in combat because of this dumb rule.

The industry missile salesmen were very persuasive and over our loudest outcries our leaders removed the "heavy, outdated guns" from our aircraft. This, when missiles did not arm for thousands of feet and launch parameters were based on unaccelerated flight. Hence, when I finally went to war in Vietnam, it was in an F-4D with no gun. Incidentally, this argument is going on again with the F-22 and follow-on aircraft. The "smart guys" want to get rid of the gun again. No way, I hope!

Air-to-air missiles came in two basic types: infra-red and radar guided. In addition, the F-101 was equipped with the Genie missile that had a nuclear warhead. The Genie, fortunately, never had to be used. The infra-red missiles had a range of about two miles (the range varied with launch speed, target aspect, target speed, and altitude). They had to be fired from the tail aspect to detect the heat from the target's tailpipe. They were limited if the target was flying toward the sun, at low altitude, or in weather. First used by Taiwanese F-86s against Chinese Communist Forces (CCF) Migs, they were very effective within parameters, but were subject to counter measures (flares, other heat sources on the ground, and the sun).[23] Launch had to be made while the shooter was below two "Gs" in accelerated flight.

There were two versions that were later used in Vietnam: the AIM-9B Sidewinder and the AIM-4 Falcon. The major differences in the two were that the AIM-4 was slightly more complex to launch (it had a cooled seeker head which made it more sensitive), and it could be launched at higher G, but as it had no proximity fuse the AIM-4 had to score a direct hit to achieve a kill. On the other hand, the AIM-9 had a proximity fuse and did not need a direct hit to kill. However, it did need 1000-2500 feet to arm and guide.

In Vietnam combat from 1965 to 1968, the AIM-9 achieved 28 kills for 175 launches or 16.9% while the AIM-4 had 4 kills for 43 launches for a kill probability (Pk) of 10.7%. That was a far cry from the 70-80% Pks predicted by the so-called experts before the shooting started![24]

The radar guided missiles were typified by the AIM-7D, Sparrow. It had about a 25 mile range, a speed of about 3.7 mach, and was a beam rider. It rode the CW Wave of the fighter's radar and homed on the radar energy reflected by the target. This required the attacker to illuminate the target aircraft with radar energy from launch to detonation.[25] To manage this in a hard "G" maneuvering fight was at best difficult. In addition, the missile did not arm until it had undergone a finite amount of acceleration and it would not guide

for about a half mile during its boost phase. In other words, inside of a half mile range it was an unguided rocket.[26] Out of 224 launches in Vietnam combat, the Sparrow achieved 20 victories for a Pk of 8.9%.[27]

Several kills were lost when the proximity fuse on the AIM-7Ds prematurely detonated. It was discovered that the fuse would detonate on a harmonic reflected from the Mig-21's engine turbine when the missile was too far away to get a kill. Typically a good missile would launch, lock-on, and guide right to the Mig. The Mig would then disappear in the Sparrow explosion only to fly out the other side unscathed. Analysis indicated the Sparrows were detonating prematurely on the engine harmonic.

Note that the F-4s in Vietnam, without guns, had no weapons system that was effective inside of about 2000 feet in range. Nor could any of

these missiles reliably track targets at low altitude. As the F-4 was the primary U.S. air-to-air fighter, one can readily see that the F-4 (used by both the USAF & the USN) was severely handicapped in this role by these missile limitations. It also appears difficult to justify the no-gun decision. Couple this with the U.S. rules of engagement that required visual identification (VID) before firing was permitted and the situation approached the ridiculous. When North Vietnamese surface to air missiles (SAMs) forced U.S. forces below 10,000 feet (the aircraft could out-maneuver the SA-2 Soviet SAM missile at these lower altitudes), it reduced the effectiveness of the U.S. air-to-air missiles as low altitude infra-red sources and radar reflection sometimes would deflect the missiles.

Another key development between the Korean and Vietnamese Wars

was mid-air refueling. Mid-air refueling was first tried by the U.S. Air Service in the 1920s.[28] It was first used in the jet age by the British, refueling Meteor jet fighters from a Lancaster bomber adapted as a tanker. USAF Colonel Dave Schilling got the opportunity to refuel during a conference with the British. He immediately approached General Vandenberg, the CSAF, and got approval to modify several F-84 aircraft for refueling. In September of 1950, he then demonstrated the concept, in spite of atrocious weather, by successfully flying the Atlantic east to west (against prevailing winds). The actual flight took more than 10 hours, covered over 3,300 miles and ended with a weather divert to Maine when his destination in New York was socked in.[29]

The method used was called "probe and drogue". A refueling hose was trailed from the tanker aircraft with a large funnel (drogue) on the end of the hose. The refueler had a probe which was three to four feet long protuding from its wing or fuselage. The probe had a valve that opened when its sleeve was compressed against the drogue receptacle inside the trailing drogue (funnel). The flying was tricky and required great care and considerable skill. In addition, equipment design was complex as the refueling hose tension had to be maintained or the hose would oscillate and whip when contact was made. Pumping systems were equally complex and control dicey as large amounts of fuel were rapidly transfered and aircraft centers of gravity shifted as the receiver took on fuel. The whole procedure was freudian in its simulation and the subject of many raunchy analogies by the pilots.

Probe and drogue refueling was standard for all tactical fighters by the late 1950's. Fighters routinely flew both oceans during the 1960s. Many of the fighter squadrons that deployed to Vietnam flew directly from the U.S. using refueling over the various islands along the way (in case of emergency diversion).[30] In any case, Vietnam was the first major war to use mid-air refueling. We used it both going to and coming home from the target. There were tanker orbits over Laos and the Gulf of Tonkin, and refueling was absolutely crucial to our operations.

As an aside, all during my Air force career, I have been curious why Americans and their NATO Allies never seem to plan for the possible use of air refueling by our enemies. I first encountered this myopia in an air defense exercise at Squadron Officer School in 1957. The rules of the exercise were that the Soviets had an air refueling capability (so bombers could be refueled). As the distaff commander (acting Soviet Chief), I refueled fighters along with the bombers and escorted the Soviet bombers with fighters. U.S. air defense aircraft had no fighter-vs.-fighter weapons (no guns or fighter air combat training). They suffered dramatically from the

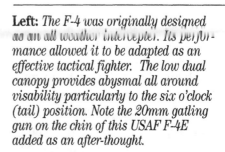

Left: *The F-4 was originally designed as an all weather interceptor. Its performance allowed it to be adapted as an effective tactical fighter. The low dual canopy provides abysmal all around visability particularly to the six o'clock (tail) position. Note the 20mm gatling gun on the chin of this USAF F-4E added as an after-thought.*

escorting Soviet fighters exactly as I planned. By the rules of the exercise I did very well, but at the critique, I was told I did not have the "big picture" and that the Soviets could not refuel well enough to conduct the operation I had planned. When I asked how the Soviet bombers had refueled, I was dismissed. (Without refueling for the Soviet bombers, the exercise was impossible and it was no more difficult to refuel fighters than bombers.)

This predilection continued in real life, both in the Persian Gulf and Vietnam. I keep wondering why we never seem to give consideration to our opponents having a refueling capability. Nor do we give adequate consideration to our tanker capability being attacked. For example, if the British had destroyed the Argentine refueling capability (only two C-130s) in the Falkland Islands, they would have had a much easier time. The Falklands were out of range for most of the Argentine fighter-bombers without refueling. Likewise, if the North Vietnamese had targeted our tankers in Vietnam, they could have precipitated a huge disaster and our operations would have been severely restricted. The same was true in the Persian Gulf.

We have been fortunate up to now. A war is coming when the "bad guys" will not only have a refueling capability, they will use it. In my opinion, air refueling has never been exploited to its fullest. For example, in an air defense campaign, why not scramble defending fighters to tankers prior to early warning, which would make them impervious to ground strikes, let

Right: *The Vought A-7 Corsair II ("SLUF"–Short Little Ugly Fellow) was like the A-4 a "Point" design. The A-7D pictured was my personal aircraft and had many new, modern innovations: a heads up display, an automatic bombing system (that increased accuracy by a factor of 2 or 3). Pencil beam air to ground radar etc. The .9 design could carry 15,000#s of ordinance over a 1000 mile radius at .85–.9 mach.*

our SAMs free fire at incoming enemy raids, then attack and intercept in mass using full tanks and supersonic speeds — sort of a "Big Wing" via tankers concept.

From the earliest days, various types of tanker aircraft were used: KC-97s, KB-50s, AJ-3s, A3Ds, KC-130s, KA-6s, KC-135s and today the KC-10. "Buddy refueling" by using another fighter as a tanker has also been used.

The Strategic Air Command developed boom refueling for SAC bombers concurrently with the probe and drogue methods used for tac fighters. The boom method was developed as the big bombers had to take on much larger offloads of fuel than the fighters, while their pilots juggled multiple throttles. In boom refueling, the tanker trailed a stiff hard boom about 20-30 feet long from the underside of tail of the tanker with controlling wings on the boom. The valve mechanism for the fuel was on the boom. The boom operator ("boomer") would fly the boom nozzle into a receptacle on the receiver. In freudian terms, with the boom method the tanker became the male, the receiver the female – just the opposite from the probe and drogue method. The flex-

ible hoses of the drogues could not equal the fuel transfer rates of the hard boom. For the heavy bombers, the boom method proved superior to probe and drogue.

In recent years, the USAF standardized on boom refueling for both bombers and fighters. Boom refueling has the advantage of being a little easier for the fighter pilot, but fighters must refuel one at a time. Pure drogue refuelers, on the other hand, can fuel three fighters at a time. The Navy still uses the probe and drogue method. Since tanker support is limited, having two systems can make scheduling tankers a problem. When refueling with the KC-135, care had to be taken to schedule the right kind of tankers with the right kind of receiver. Fortunately, boom refuelers could be converted to drogue refuelers, but it had to be done before the fact on the ground. This definitely affected operations in Vietnam and later in the Persian Gulf. The KC-10 tanker, which came along after

Vietnam was over, incorporates both refueling systems directly on the aircraft. The KC-10 can refuel Navy aircraft and USAF aircraft back to back without cumbersome ground re-configurations.

VISITS TO VIETNAM

In December of 1959, I visited Saigon enroute to Bangkok, Thailand. At that time, the U.S. had a small military assistance group (MAAG) operating in Saigon. The Viet Cong (VC) and North Vietnamese Army (NVA) were supporting a terrorist campaign against the leaders of the Diem government and the governmental infrastructure. Several junior officers (contemporaries) I talked to were very worried about the effect of this campaign on Diem's government. They also told me there was a shooting war going on in Laos. Saigon in 1959 was a beautiful city. I remember the food in the restaurant where I ate as being delicious (a combination of French and Oriental). In 1968 (after the Tet,

offensive) when I passed through again, I did not recognize the city. It was hiding behind barbed wire entanglements, and sandbags, and was visibly scarred. I had a hamburger at the BX.

My friends' misgivings in 1959 proved well-founded. By late 1960, Ho Chi Minh's "Armed Revolt" had won large portions of the Republic of Vietnam (RVN) and many government officials had been assassinated.[31] In November of 1960, there was an attempt to overthrow Diem. In 1963, with the approval, if not the total complicity of the U.S., Diem was overthrown in a military coup which tragically resulted in his death."

There followed a series of coups and the U.S. became entangled deeply in South East Asia.

THE TONKIN GULF

On July 31, 1964, a U.S. Navy destroyer, the USS Maddox, began an intelligence-gathering patrol in the Gulf of

Tonkin off the coast of North Vietnam. On 2 August, three North Vietnamese torpedo boats closed with the Maddox at high speed. In a confused battle, one torpedo boat was sunk and the other two damaged. The Maddox was assisted by aircraft from the USS Ticonderoga. The U.S. protested to Hanoi and warned of "grave consequences".[33]

On August 4, 1964, the Maddox accompanied by the USS Turner Joy, again encountered torpedo boats. Again two torpedo boats were sunk and two damaged. The next day, aircraft from the Constellation and the Ticonderoga launched retaliatory strikes against North Vietnamese naval bases at Vinh. The strikes were successful but anti-aircraft artillery (AAA) fire was heavy and the U.S. had two aircraft shot down.[34] Two days later 30 Mig-15 and Mig-17 aircraft reinforced the NVA forces at Phuc Yen Airfield. That very same day, Congress passed the Gulf of Tonkin resolution which authorized the President "to take all necessary steps" to meet the emergency. Two squadrons of U.S. B-57s deployed into Bien Hoa outside Saigon while several squadrons of F-100s and F-102s landed at Da Nang. F-105s were deployed to Takli Air Base in Thailand.[35] The die was cast.

THE VIETNAMESE AIR WAR

As air wars go, the Vietnam War was an anomaly. It began by the USAF and USN air attacks in response to North Vietnam and Viet Cong escalations, both in South Vietnam and the original attacks in the Tonkin gulf. Each time a VC/NVA attack was made on one of the southern bases, we would reply with a raid up north.[36] Air Power was not employed in traditional, coordinated attacks, but as punishments for enemy transgressions and provocations. It was an uninformed way to use our fighter air power and wasted lives and machines. Its success was questionable at best.

Vietnam was the first test of our second and third generation jet aircraft. Many new systems were developed and some old peacetime myths blown away. The new systems that first appeared in combat in Vietnam between 1964 and 1975 were: supersonic aircraft; surface to air missiles (SAMs); air to air missiles (AAMs); "Smart" bombs (laser guided and electro-optical bombs); air refueling was used on a massive scale; airborne communi-

cations & control centers (ABCCCs) were used for the first time; C-47, C-119 and C-130 Gunships; electronic warfare became a science; the use of all sorts of new sensors including low-light TV, starlight scopes, infrared, acoustic, and seismic were pioneered; and the massive use of helicopters in every type mission. The list over the 11 major years of the war is almost endless.[37]

Three myths were quickly dispelled:

1) "Supersonic aircraft were too fast to dogfight". We have heard the old saw that fighters are too fast to dogfight before each and every war since WW I. We are currently hearing it again. Before WW II the British adopted the three-ship Vic formation because the aircraft were so fast that dogfights were considered not possible. In Korea, the jets were again allegedly too fast to fight air-to-air.

In Vietnam, it was the NVAF Migs who remembered Korean tactics. Equipped with a fleet of fighters that were technically supersonic, the USAF & USN forgot that when they carry bombs externally they are subsonic bombers. We also forgot that when supersonic aircraft maneuver

they decelerate to subsonic speeds very quickly and to get supersonic requires afterburners full-on and gobs of fuel. Subsonic NVAF aircraft, under good radar ground control, proved to have a real intercept capability against our supposedly superior aircraft.

Finally, we were restrained politically from exploiting the superior qualities of our weapons systems such as medium-range air to air missiles.

2) "Fighter-bombers are too fast for Anti-aircraft artillery (AAA or Flak)". Common sense should have put this one to bed. Flak can be pointed far enough ahead to lead any fast mover and barrage fire can be very effective at low altitude when aircraft fly through the barrage. Yet years after Vietnam in 1977, I had a knockdown, dragout argument with a British Air Vice Marshall in NATO about the efficacy of light flak. He still thought modern aircraft were too fast to be hit. Having been tracked at near sonic speeds at 100 'AGL; I knew he was wrong. He would not listen. British Tornados later took losses trying to bomb at low level in Iraq during Desert

Left: *The F-104 exceeded mach 2 in 1954, but its high tail and small wing limited maneuverability. Shown is the two-seater F-104B.*

Storm! When I checked into Desert Storm with the big boys who were there, I found the USAF in Desert Storm had moved up and out of small arms range (10,000/15,000 feet) the minute our ECM experts, and missile fighters ("the Wizards and Wild Weasels") had limited the missile threat and we were not losing multi-million dollar aircraft to light Flak. In fighter tactics, stupidity will happen, if you let it.

3) "Fighters do not need guns for close-in air-to-air combat". We have already discussed the missile versus the gun and it is obvious they are supplemental. The gun still allows you to kill the other guy when eyeball to eyeball. It also is very useful against soft ground targets and troops. Remember, 50% of the casualties in WW II caused by aircraft were caused by strafing." If you use a 30 MM such as mounted in the A-10, it works for armor too.[39]

ROLLING ("GRADUATED")[40] THUNDER

On February 13, 1965, President Johnson approved Operation Rolling Thunder. The basis of this operation was found in a National Security Action Memo (NSAM #288) from the year before where the President outlined the political objectives of an air campaign against the North. In NSAM #288, Secretary of Defense (SecDef.) McNamara directed the Joint Chiefs Of Staff (JCS) to develop a program "of graduated overt military pressure" against North Vietnam. McNamara also ordered a more limited program of retaliatory raids that could be initiated on 72 hours notice. Instead of replying to the SecDef that a graduated response doomed us to a long war of attrition against a determined enemy, the JCS responded with an Operations Plan that linked retaliatory raids with continuous bombing of steadily increasing intensity and which could still meet the SecDef's requirements. The plan called for attacks on the full gamut of North Vietnamese targets and included mining all of the North Vietnamese harbors.[41]

None of the President's civilian advisors recommended that the President support this JCS plan (which was very similar to Linebacker II which was unilaterally implemented by President Nixon in 1972). When questioned about the objectives of their plan by the SecDef, the JCS replied that if he or the President would define a specific political objective then they would define a specific military objective and a course of action to achieve that objective.[42]

The problem in Vietnam was that none of the President's civilian advisors, nor the military, nor even the President himself, could agree on any of the objectives. A civilian committee headed by the President selected targets each week for Rolling Thunder. For a time the JCS was not even invited. Presidential advisors tried to pick targets, the destruction of which, would be so painful that the North Vietnamese would stop supporting the war in South Vietnam. At the same time no actions were allowed to be taken by our military that could possibly result in any escalation that might bring Communist China or Russia actively into the war. Their greatest fear was to commit some act that might remotely raise the possibility of nuclear escalation by the Communists. Nor was President Johnson interested in mobilizing the full might of the U.S. and disrupting his plans for a "Great Society" at home.[43] The moral dichotomy of building a "Great Society" while other Americans were dying in the mud 10,000 miles away did not appear to occur to him. But it was readily obvious to the veterans who came home to the "world" to find draft-dodgers being recognized as heroes, and business as usual, everywhere except in Vietnam.

A clear, obtainable National Objective was never articulated in our Vietnam policy. The national objective therefore, devolved by default, into a vague, graduated attempt to destroy the will of the North Vietnamese Communists to continue their insurrection in the South. Most of the Communists had dedicated their lives to reuniting Vietnam. The U.S tried to defeat these dedicated men by graduated air strikes that had little effect on them. Secretary McNamara actually said this in 1965; "we should try to destroy the will of the DRV to continue their political interference and guerilla activity. We should try to induce them to get out of the war without having their country destroyed and if they do not get out, their country will be destroyed."[44]

In this situation, the fighter pilots in combat had to sort this disparity of clear objectives, striking nonsensical targets with ineffective weapons, while risking their own lives under fire -- Operation Rolling Thunder. Or was it as Clodfelter calls it "Graduated Thunder."

The Operation began on 2 March 1965 when the South Vietnamese (19 A-1s) and the USAF (44 F-105s, 40 F-100s, and 20 B-57s) struck Quang Khe. Four aircraft were

lost to AAA. These losses caused the USAF to reevaluate its tactics. Low-level attacks below 4,500 feet were discarded to get out of most of the light flak. Only one pass was to be made and everyone was to randomly "jink" (use random turns and maneuvers in pulling off the targets) to destroy gun tracking solutions.[45]

Initially, Rolling Thunder was limited to targets below the 20th Parallel. Targets were restricted to those facilities directly associated with the flow of supplies; i.e., roads, bridges, barracks, depots, rail lines, and choke points only. Attacking ports, harbors, and airfields was strictly prohibited.[46]

Migs first attacked USN aircraft who were bombing the Than Hoa bridge on April 3, 1965. There were no losses on either side. The next day, two F-105s were lost to radar-controlled Mig-17s while attacking the same bridge (the leader of that flight was a friend of mine and a damn good fighter pilot). So much for the lack of a dogfight capability in the F-105 (a nuclear penetrator) being irrelevant.[47] As a result of this fight, the F-100 was withdrawn (as an air-to-air escort to protect from enemy fighters) from use in North Vietnam and was replaced by the F-4 Phantom.[48]

The U.S. also had to reevaluate the effects of flak on supersonic aircraft. By mid-1965, 50 aircraft had been lost to flak. In the first two years of the war 85% of the 455 aircraft lost, fell to flak. The DRV had over 5,000 guns emplaced, half around Hanoi and Haiphong. These included 12.7 MM, 14.5 MM

ZPUs, 23 MM, 37 MM, 57 MM, 85 MM, and 100 MM – plus all the AK-47s, rifles, and machine guns that were used by the NVA in barrage fire (basic Soviet AAA doctrine). The rifle caliber guns, 12.7 MM and 14.5 MM, of which there were many, ran out of accuracy above 4,000 feet. The 23 MM was accurate to about 6,000 feet, the 37 MM to about 7,000, the 57 MM to 15,000 and the heavy guns to approximately 20,000-25,000.[49]

When the SA-2 SAM appeared, it forced the fighters back down to low altitude where even rifle caliber weapons were effective. The first successful SAM engagement

occurred on July 24, 1965. An SA-2 destroyed an F-4C and damaged the three others in the flight who were close by. The fighters could out-maneuver single missiles if at low altitude and in visual contact. This limited operations to clear weather.[50] To fly in weather in a SAM area was death.

To outmaneuver the missile, the classic maneuver was to pushover in a dive (using negative Gs) and get the missile headed down and then to execute a maximum G pullup. Below 15,000 feet, the SA-2 could not reverse back upward and follow the fighter. Above 15,000 feet the SA-2 had more available G for maneuvering than the fighter. If the pilot was in the clouds and unable to see or he did not spot the missile early enough to maneuver, he was in trouble. If the SAMs were fired in clusters, it became much more difficult to avoid them (and to control them).[51]

The SA-2 was a beam-rider type radar missile that was later modified to be optically launched. U.S. reconnaissance picked up on the construction of the first sites, but Washington prohibited striking them for fear of injuring or killing the Russian technicians obviously involved. The Missile used general guidance from the tracking radar to guide for its initial launch parameters then went to a high pulse repetition frequency signal once it was airborne and for its final guidance. The missile did not guide during its initial rocket powered acceleration phase. Since all of these radar signals could be picked up by our receivers, the war entered an ECM phase that has been a technological trend ever since.[52]

Below: *The General Dynamics F-111 should have been called the "B-111" according to many fighter pilots. Regardless at mach 2.5 it was the fastest USAF operational fighter.*

The first step was the installation of Radar Homing And Warning receivers (RHAW) in the fighters. As the name indicates, this was a passive receiver which provided very basic analog information as to radar threats: ground radar, airborne intercept radar, and SA-2 radars. The SA-2 signals were broken down by general activity, launch, tracking and high PRF lights and tones. All signals were presented on a circular dial that showed azimuth and rough range information (based on the strength of the signal). These visual signals were augmented by audio tones that were readily discernible. The high PRF signal associated with final guidance of the SA-2 sounded like a rattlesnake (ECM wizards have a perverse sense of humor). This equipment saved lives as the heads-up gave the pilot time to maneuver. In addition, in many cases pilots spotted missiles previously unseen because of the RHAW gear.[53]

After the first losses to SAMs in 1965, the USAF introduced EB-66B radar jammers and F-100F "Wild Weasels" (Wild Weasel was the nickname given the aircraft tasked with finding and attacking SAMs). The EB-66Bs were standoff jammers and blinded the radars for AAA and SAMs. The F100Fs had equipment installed to locate the missile radar and guidance systems and accompanying fighters would attack the missile sites. Unfortunately, Washington was still worried about Soviet casualties, and initially limited the number of strikes to just a few SAM sites. Later on, the Pentagon required analyzed photography of the target-

ed site before a strike could be made. Invariably, the NVA would relocate the SAMs and their vans before the SAMs could be hit.[54] It was "Catch 22".

When the F-100F was fitted with the AGM-45A Shrike missile the game changed again. The Shrike was a radar homing missile that followed the radar beam of the SA-2 Fansong acquisition radar and destroyed the radar antenna and van.[55]

The F-100F was never satisfactory as a Wild Weasel aircraft as it had less range and less speed than the F-105Ds it was supporting. In the summer of 1966, the F-105G was introduced as the primary Weasel aircraft. Fitted with Shrikes and improved radar detection gear, this two seater was the bane of SAM operators until the end of the war.[56]

Faced with a Shrike, the SAM operator had to shut his radar down or face possible destruction. Unfortu-

nately for the Weasels, the Shrike launch could be seen by an experienced operator on the tracking radar. Once the radar was shut down, the AGM-45A went ballistic and had no further guidance capability. But to get the radar to shut down was as effective as attacking the site as it protected the strike force which was the Weasel's mission.[57] By 1968, each fighter was fitted with its own ECM pod to blind the radars used by AAA and the SAMs. These pods when used in line abreast formation were very effective.

Rolling Thunder continued from 1964-1968 and the electronic war continued to escalate. The F-105 had taken over as the primary bomber and the F-4 as the air-to-air protector; however, as the war progressed, the F-4 was used as a fighter bomber as well.[58] Actually the F-4 had more bomb-carrying capacity and range than the F-105. The author flew many missions in

F-4s in Laos and southern North Vietnam with eighteen 500 pound bombs (6 more than a max load for a B-17 in WW II).

In addition, there were clandestine interdiction and close support operations during these years in Laos. Northern Laos was known as "Barrel Roll" and southern as "Steel Tiger". All of these operations were air-to-ground and never discussed publicly. The Migs and SAMs through most of the war flew only in protection of North Vietnam (that changed when the DRV invaded South Vietnam with conventional force in 1972). Because of the fluid situation on the ground these missions in Laos were almost always directed by a Forward Air Controller (FAC). Thousands of sorties (a sortie is one mission by one aircraft) were flown in these areas. The DRV main supply line, the Ho Chi Minh Trail, came down through mountain passes south from North Viet-

nam into Laos, then down Laotian roads to supply points just west of South Vietnam, then east into South Vietnam. During the war, they were bombed most of the way.

These supply routes were defended by light and heavy flak. The DRV mountain passes, Mu Ghia and Ban Ka Rai, had 85 MM and 100 MM guns that were radar guided; however, the most prevalent guns along the trail were 37 MM and 23 MM. Heavy machine guns and automatic rifle fire were everywhere and NVA doctrine was to fire every available weapon in likely area barrages rather than use aimed fire. Below 4,500 feet both the 14.5 MM ZPU and 12.7 MM (.50 caliber) could be serious. Interestingly, the gunners usually left the slow-moving FACs alone, probably because the FACs could call in the entire Air Force if the gunners missed. Some places in Laos were much more heavily defended than parts of North Vietnam,[59] but when 100 missions were counted for early

rotation, only the North counted. This lead to a clamor to get across the border at all costs by the aircrew. Later the policy was changed to one calendar year regardless of the number of missions.

During the numerous bombing halts, (there were eight altogether)[60] the sorties into Laos continued. This interdiction effort was not successful as long as the NVA and Viet Cong (VC) were conducting guerilla operations which required very little day-to-day supply. This all changed in 1972 when the NVA went over to conventional army attacks that needed thousands of tons of resupply each day. To move these amounts of resupply, the NVA had to move by vehicle in the daytime and clear weather. These were corps size offensives where the NVA used hundreds of tanks, artillery pieces and vehicles and their supply lines were exposed to our air power. Our air had a field day. With this air supprt from the U.S., the South Vietnamese Army soundly defeated the NVA. Many Americans today are not aware of, or have forgotten this resounding success by the South Vietnamese.

Fighting a Guerilla War, the VC could pick their time to conduct operations until they had enough supplies. The VC guerilla operations needed about 380 tons of supplies per day, but of that number, only 34 tons/day were

needed from outside sources. In other words less than one percent of the supplies landing in the North needed to be transported to the south. This small amount was impossible to stop.[61]

I participated in this interdiction war in Laos and southern North Vietnam. Project "Igloo White" was a unique effort to interdict these routes from north to south electronically. The 25 Tactical Fighter Squadron, of which I was a member, flew specially equipped F-4s to implant an electronic barrier across Laos and the Panhandle of North Vietnam.

To do this, our brand-new F-4Ds were equipped with special LORAN "D" navigational equipment, along with acoustical and seismic sensors that were dropped from our aircraft. These sensors

Below: *Mig-17s under radar control shot down two heavily-laden F-105s in their first encounter proving supersonic fighters carrying external bombs were "subsonic bombers". Mig-17s could out maneuver and out accelerate the Thuds because of high power to weight ratios.*

would be implanted in the ground (seismic) or would hang in trees (acoustical) and when a truck drove by, the sensor would send a radio signal to an aircraft circling at high altitude. This aircraft relayed the data to a computer center located at Nakom Phanom (NKP) in Thailand. Our F-4s also carried KB-18A panoramic cameras which identified precisely where the sensors were located. Using this data, the NKP center could plot the location of trucks or supply vehicles in real time. The supplies then could be attacked directly or the roads and areas in front of them could be mined or disrupted. For sure, this was a new mission for a Mach 2 fighter, but there you are.

As the sensors were delivered from a level flight attitude at low altitude, sensor droppers got nervous when the targeted areas were heavily defended. Altitudes, speed, and heading had to be exact and as the sensors ejected to the rear from their dispensers, one at a time, drop runs could be as long as two minutes straight and level. Speed usually was exactly 500 knots. When tracers were curving right over the canopy, it was nervous time.

I would like to convey that the operation was a smashing success, but, although it did have good results occasionally, its returns do not now seem to have justified the huge cost. We can report one concrete accomplishment however. A tape from an accubuoy (acoustical sensor) recorded the breaking of a tree limb and an NVA soldier falling from the tree. He was trying to get the fabric from the accubuoy's parachute. The fall broke his neck. NKP dutifully credited us with one confirmed NVA soldier as KBA (killed by aircraft).

In country, the war devolved mostly into close support missions. In general, the AA defenses were not as heavy as in Laos or North Vietnam, but the payback could be much higher. Working with friendlies requires first-rate coordination and iron discipline to avoid "friendly fire" incidents and successfully delivered air support was of great help to the "grunts". Many aircraft were used in-country that could not be used in the heavily defended North. For example the F-100s, when withdrawn from up North, were successfully used in the South. The South Vietnamese Air Force used T-28s, A-1s,

AT-37s, and F-5s. There were also several wings of USAF F-4s used. The air and artillery superiority of the friendlies forced the NVA and VC to try to closely engage our forces so this superiority could be minimized. The NVA/VC never really overcame this disadvantage in firepower, but made enormous propaganda gains from the collateral damage inevitable with the use of these powerful area weapons.

THE AIR SUPERIORITY WAR

By the end of 1965, the North Vietnamese introduced the Mig-21 to counter the F-4s now swarming over North Vietnam. The USAF and USN were never allowed to go all-out for air superiority as we had in every other war. Consequently, the Migs could attack us using their ground radar controlled intercept (GCI) capability at anytime they chose.[62]

The real difficulty in our air war was the rules of engagement which required a visual ID (VID) by U.S. aircraft of any bandits before being cleared to fire. If anything was learned in the Vietnam War, it was the importance of identification of friend and foe (IFF) in the air.

To VID a NVAF aircraft, meant getting close to the Mig and usually close maneuvering combat where a gun was required, and our F4B, C, D, and J had none. Even more importantly, our Sparrow and Sidewinder were useless inside a range of 2,000 feet.[63] In 1968, the F-4E was introduced with an internally installed 20 MM Gatling gun in

the nose. The F-4Es got seven Migs with the gun.[64]

In 1967, Colonel Robin Olds was able to coordinate a single air operation where our missiles could be employed the way they were designed to be used. Operation Bolo was created to entrap the NVAF by simulating an F-105 strike with F-4s. The area was sanitized of all other friendly fighters so no VID was required before shooting with our longer range missiles. The results were gratifying. Seven Migs were shotdown and we lost no aircraft.[65] It made clear once more that in an air superiority fight IFF is a vital requirement and the best minds in our country should be constantly working that problem.

As the air war progressed the NVAF became more and more sophisticated and the Mig-17, 19, and 21 were more maneuverable than our F-4s and all the Migs had missiles and guns. NVAF ground controllers were becoming more and more adept at high speed sneak attacks using both missiles and guns.

We were suffering from a lack of air-to-air training and the lack of a gun on some models of the F-4s. The Navy never did get a gun on their F-4s. Some of the U.S.Navy aces were quite open about their feelings in not

Above: *Powered by a large after burning Pratt & Whitney J-75 the Convair F-106 was a bonafide mach 2 intercepter with a good air to air capability. Late models were re-equipped with 20mm Gatling guns.*

having a gun. Several thought they could have substantially increased their scores had they had one.

By 1968, they had become formidable. Consider the following:

AIR TO AIR 1965-1968

	VICTORIES	LOSSES
• USAF	86.0	41
F4C/D	58.5	
F105D/F	27.5	
• USN	34	14
F-8/C/E/J	18	
F-4B/J	14	
A-1E	1	
A-4C	1[66]	

From 1900 to 1971, the USAF ground troops were being withdrawn from Vietnam and the period was marked by a pronounced NVA buildup of armor, artillery, and conventional forces at all of her supply points in southern North Vietnam and Laos. Except for reconnaissance, USAF missions and air activity stopped over the North.[67]

When "protective reaction strikes" (missions to protect our reconnaissance birds) began in January 1972, the air-to-air war flamed to a new intensity. In the interim, the USAF and particularly the USN, had begun to retrain its aircrew for air-to-air combat and the difference was immediately apparent. The U.S. had incorporated better airborne radar over the North from radar picket aircraft and navy ships so that VID was required less frequently.[68]

In addition, as Linebacker I bombing began in earnest in April 1972, the brunt of the bombing, Wild Weasel, and air-to-air missions were taken by the F-4 as the F-105 had been phased out (except for some F-105 Wild Weasels). These F-4s were potent air-to-air vehicles when the bombs were gone. Also the introduction of pinpoint bombing by laser guided and electro-optical "smart bombs", made previously impervious interdiction targets (such as the Paul Doumier and Dragon bridges) easier prey.[69]

In May, the USAF and Navy shot down 27 Migs, but the fast Mig-21 was still formidable. In July, two F4Es were on Mig Cap and caught, as so many more before them, by surprise from supersonic Migs zooming up under radar control from an undercast below. The calls to break came too late. Even the guns in their noses proved no defense to rapid attacks unseen from six o'clock low. The USAF retaliated on July 8 by shooting down three Mig-21s. These air-to-air battles continued until the end of October with the USAF and USN gaining ascendancy. When the NVAF introduced the Chinese version of the Mig-19, the F-6, several were shot down. These activities ceased as the U.S. and DRV sat down to "earnest" peace talks in October.[70]

LINEBACKER I

When the NVA attacked South Vietnam in 1972, they struck from three directions with major forces: across the DMZ, in the Central Highlands, and out of Cambodia towards An Loc. U.S. Intelligence had predicted the offensive to begin in January. By waiting until March, Giap gained surprise and the weight of the three major, simultaneous attacks by the NVA was unexpected. Almost all of the American ground troops were out of Vietnam. The NVA hoped to overwhelm the ARVN and push the Americans out for good. To emphasize the point, the NVA opened a fourth offensive in Binh Dinh province along the coast.

SA-2 missiles were moved into the De-militarized Zone, Laos and the northern part of South Vietnam. For the first time aircraft in the South were attacked by missile weapons previously only faced up north. The SA-7, a shoulder-launched IR missile was introduced for the first time. It was deadly to "slow movers": Forward Air Controllers (FACs) flying O-1s, O-2s, and Ov-10s, helicopters, and C-130 gunships; all of whom took losses.

President Nixon was determined that the South Vietnamese would hold and Air Force units from around the world reinforced the units already there. American close air support and ARVN tenacity saved the day.[71]

Right: *A little known program in Vietnam was called "Buffalo Hunter" which used drone aircraft for reconnaisance and ECM jamming. Pictured is a Ryan "Firebee" Drone used in this program.*

LINEBACKER II

President Nixon, with characteristic courage, sent waves of B-52s to Hanoi on December 18 to begin Linebacker II, or as the USAF pilots called it "The Twelve Days of Christmas". Over the 11 missions (none on Christmas Day) flown from 18-29 December 1972, 729 B-52 sorties were flown, 1,600 military structures, 372 railroad cars and locomotives, and 3,000,000 gallons of POL were destroyed. All of the DRV's airfields, industrial areas, roads, railroads, and networks were heavily hit. Over 1,240 NVA SA-2s were fired and there were few left. Electrical generating capacity fell from 115,000 kilowatts, to 29,000. Rail traffic within 10 miles of Hanoi was totally disrupted.

Over 15,237 tons of ordnance was dropped north of the 20th parallel. About 5,000 tons of this total was dropped by fighter-bombers. A-7s from my wing destroyed Hanoi Radio during this period in downtown Hanoi. Over 100 bombs struck the station or the small yard surrounding it. The destruction was total. No A-7s were lost.

The DRV did not expect this all-out response and the magnitude of the B-52 raids devastated them. The DRV was prostrate and asked for peace on December 30. The cost was 18 B-52s lost.[75] The Armistice

Linebacker was initiated in April 1972 to blunt the offensive by the NVA. The U.S. Navy promptly mined the harbors along the Vietnam coast. President Nixon, by brilliantly opening diplomatic relations with China and sponsoring detente with Russia had quieted the threat of the war widening to the communist major powers and there was a very mild reaction to our mining the harbors[72] (Nixon is rarely given credit for getting rid of this threat of escalation which according to McNamara was his biggest fear. Perhaps Nixon was a little better and brighter than the "Best and the Brightest".)

The sizeable conventional attacks of 150,000 men equipped with T-54, T-34, and PT-76 Tanks demanded a very high level of resupply and USAF and USN close support and interdiction missions began to take a terrific toll of the NVA troops, vehicles, tanks, and resupply. The DRV was not resupplying guerillas now, but a large army in the field and the effectiveness of the USAF and USN airpower was multiplied on an

exponential scale. This time the DRV had bitten off more than it could chew and the Communists asked for negotiations in Paris. The DRV needed to get the Americans out of Vietnam and our airpower off their back. The Americans wanted their prisoners of war back and their military totally out of Vietnam. Linebacker I ended on 22 October when a cease fire was called. It was obvious to Secretary of State Kissinger that the DRV wanted peace. He stated peace was at hand on October 26.[73]

Negotiations broke down again on November 23. On December 13, the DRV representatives walked out of the peace talks. President Nixon finally decided to use strategic bombers strategically by ordering the B-52s to go "downtown" to Hanoi and Halphong. Prior to November 1972, we had bombed North Vietnam with tactical fighter-bombers and attack aircraft only. On November 22, 1972, B-52s were been used over the north and a B-52 was shot down by a SA-2 at Vinh. This was the first B-52 lost in the Vietnam War.[74]

was signed on January 9, 1973.

SEARCH AND RESCUE
(SAR)

The most satisfying missions I ever flew were on search and rescue operations. In my view, these missions have never received the attention they are due. Only recently with the rescue of Scott O' Grady from Bosnia, did a SAR story and its innate heroism come to the attention of the American people. When one of our pilots went down in Vietnam, the war literally stopped until we could get the downed aircrew out. Those SAR efforts were incredible for their oganization, skill, intensity and bravery. It was awesome to see the professional skill and the total courage of the men and women involved. In addition, these missions had an enormously positive effect on aircrew morale in Vietnam.

When an aircrew went down, the first step was to locate the downed person and positively identify him on his survival radio. The airborne SAR Commander used a "Sandy" call sign. Until 1972, Sandies flew A-1s. In 1972 this mission shifted to A-7s and later, after the war to the A-10. The SAR Commander would assume control of the area where the downed aircrew was located and provide suppressing fires to protect the aircrew and clear the area of hostiles. An Airborne Combat Control Center ("Crown") would funnel necessary air resources to the SAR Commander as needed. A big SAR looked like Armed Forces Day at a big airfield in the states. Using all the resources available, the SAR Commander would totally flatten the area around the downed crewmember using area munitions

and also clear a route in for pickup by a rescue helicopter. When the site was prepared, the helicopter would be cleared-in with Sandies daisy-chaining (usually flying a figure eight) around the helicopter using area munitions to suppress any possible fire at the "chopper". No Sandy, or "Jolly Green" (Helicopter pilot/crew) will ever buy a drink in any bar that I am in, so long as I am breathing. They are the bravest men I have ever known.

Let me give one personal example. One night at Ubon Royal Thai Airbase in December 1968, I checked the fragmentary operations order ("the Frag") to see where I was going on my F-4 mission the next day. I was ordered to drop sensors on a truck park located at Tchepone in Southern Laos ("Steel Tiger") at 0600 hours (6:00 AM). Tchepone was an important NVA supply point and road junction on highway 911 running north to North Vietnam that joined route 9 running east into South Vietnam. It was a key point on the Ho Chi Minh Trail. There was also a confluence of three rivers that provided river transport by light boats or bateaux (the rivers formed an "H" which was very visible from the air and showed up clearly on aircraft radar). The NVA defended Tchepone enthusiastically with numerous 37 MM, 23 MM, 14.5 MM and 12.7 MM guns. The gunners there were experienced and aggressive. They fired at anyone who went near the place.

Since I had to directly overfly the area covered by the guns at 500 feet (all of the guns were effective at that altitude) and had to remain straight and level for about 30 sec-

onds to release the specified type of sensors, I was not all that thrilled to discover my mission. The good news was that the drop was to be made at 500 knots indicated (the speed was dictated by the sensor drop). Hopefully, I could catch them asleep and get through before they could track me.

Our squadron assigned specified "lead crews" to drop each sensors "string" (the sensors were dropped in lines along the ground that we called "strings" and as the sensors had to be replenished every two or three weeks, we often flew the same mission over and over again. This "pedigreeing" as we called it helped with the required low level navigation at high speed as we knew the checkpoints and the navigation had to be done with pinpoint accuracy. Photo interpreters had to locate exactly where the sensors were after we dropped them. Pedigreeing was an advantage except when you drew a difficult, well-defended target and had to go back again and again. The Tchepone String was assigned to me from day one and I knew the low-level approaches to Tchepone cold, but I was also aware that those gunners "hosed away" everytime we went through there. It was a "nervous" night. After a short night and a quick breakfast, I reported to operations for briefing at 0330 only to find that my mission had been scrubbed. A high-speed FAC (forward air controller flying an F-4) had been hit by a 37 MM the evening before when overflying Tchepone and knocked out of control. When he went out of control, the pilot told the "GIB" (Guy in the backseat, usually a navigator or sometimes another pilot) to eject. They were at high speed (about 600

knots) and low. The deceleration on ejection caused the GIB's legs to flail in the 600 knot slipstream and he compound fractured both his legs. The force of his ejection caused the aircraft to recover and the pilot flew home to a "hairy", but safe crash landing (he walked away, but the aircraft was junked).

The GIB was in a field right in the heart of the NVA facilities at Tchepone and with two severely broken legs had crawled under some brush and hidden in some tall grass in a rice field. He was hiding about a quarter mile east of where I was supposed to drop sensors and had been talking at hourly intervals to the Sandy commander on his survival radio. Bad guys were all around him, but did not know where he was. This was one brave "Jose", this GIB.

The reason my mission had been scrubbed was obvious and my flight of two ships was being assigned to the SAR effort. We were loaded out with rockets and cluster bomb units (CBUs) both of which do outstanding work on AAA emplacements.

As I checked in with the Sandy CO, there were scores of Air Force and Navy fighters circling Tchepone and the NVA gunners had gotten really restless and were throwing up curtains of flak. Just before we arrived, a U.S. Navy A-4 bombing the guns was shot down with no chute or survival beeper. The Sandy Commander put us in on an active 37 MM gun site west of the river that formed the western leg of the "H". This gun emplacement was just across the river from the field where the GIB was hiding.

We got good hits on the gun emplacements and they ceased firing, but several other 37 MMs and 23 MMs came up as we pulled off. With some ordnance remaining, we came around again and dropped on a suspected 23 MM site with undetermined results. With our ordnance gone, we pulled off and Sandy put some Navy F-4s in behind us. As we were leaving the target area, a Navy F-4 was hit and went in. Again, there were no chutes or beepers. Climbing out we remained on SAR frequency listening to the fight. The bad guys were getting closer to the GIB on the ground, so Sandy Lead, after some more flak suppression, called in the rescue chopper. That big slow HH-53 flew into the area I had been apprehensive about in a 500 Knot F-4. The chopper got the GIB on board as the Sandies (A-1 prop aircraft) clobbered every thing that moved. On lift-off, an NVA 37 MM no one had seen, came up and hammered the chopper. The chopper auto-rotated into a field across the river close to where my flight had attacked the 37 MM. After taking care of the gun that had hit the HH-53, Sandy lead brought in a second rescue chopper that picked up everyone including the GIB. One of

the first helicopter crew, the door gunner, was killed as he had been hit by the 37 MM fire when they had been shot down. An hour later, everyone else was safe on the ground at Ubon.

The party a Ubon that night was legend. One of ours was lost and had been found. The GIB arrived at the hospital with a blood- alcohol content that the medics had previously thought fatal. He felt nothing until the next day which must have been awful.

I really felt proud to have participated in this SAR. The Sandies Jolly Greens, and that GIB were brave men. They ennobled us.

AFTER LINEBACKER II

I returned to Thailand as an A-7 Squadron Commander in 1973 on a six month TDY (temporary duty). The A-7 was an ideal weapons system for 'Nam. It could carry 15,000 pounds of ordnance over long distances and had an automatic bombing system that was roughly twice as accurate as any other dumb bomber (the wing's dive bombing circular error average (CEA) in its operational readiness inspection prior to our squadron deployment overseas was only 39 feet). We found this accuracy even better with our peaked systems in combat. My squadron accounted for as much BDA (bomb damage assessment) as any two F-4 squadrons in South East Asia at the same time. We could carry twelve 500 pound bombs and 1,000 rounds of 20 MM from Korat, Thailand to Hanoi, spend 15 minutes over the target and return to Korat with an adequate fuel reserve, unrefueled. As a Laotian FAC said as my flight checked out in Barrel Roll after obliterating his target, "I think A-7 velly nice ailprane."

It was a different war in 1973. After the mining of the harbors up North, and Linebacker II, there was little or no enemy activity in Laos to speak of. Resistance had disappeared. Only the Khmer Rouge in Cambodia were still fighting. I came home in March of 1973 as I had been promoted to full Colonel and I was needed as Asst. Director of Operations at my home base at Davis-Monthan in Tucson, AZ.

SUMMARY

Vietnam was a national anomaly where a misunderstanding of how to use Air Power was combined with a violation of the Principle of the Objective by our national leadership. President Johnson's fear of a nuclear war with China or Russia combined with his desire to maintain the normalcy of peacetime at home, led to a gradual war of attri-

tion that the nation's soldiers and airmen were asked to endure. No clear and attainable objective was ever spelled out. As the air war was really a tactical fighter's war, the fighter pilot bore more than his share of this idiocy.

When Richard Nixon was elected President, he adroitly eliminated the danger of major war proliferation with Russia (detente) and China (diplomatic recognition). He consequently could take full measure of the North Vietnamese and struck with all the might of our air power. Clodfelter points out that his objectives at this time were very doable and were much more limited than those of President Johnson [76] With a major conventional war going on down south, there were plenty of targets suitable for our air power. Interdiction works well when conventional enemies are attacking. It does not work to stop a guerilla war.

The Vietnam War was the first war where mid-air refueling came into its own. We refueled inbound to and outbound from North Vietnam on almost every mission. Practically every USAF squadron flew to Southeast Asia across the Pacific with tankers. In 1969, my F-4D squadron flew from Eglin AFB, Florida to Vietnam with an RON at Hickam in Hawaii. Air refueling is not an American monopoly although we have come to think of it that way. In the next air war I expect our tankers to be targeted by some thinking enemy fighter pilot and the enemy will be using his refueling capability to strike at us. Lack of fighter range cost the Germans in WW II, the North Koreans/Chinese/Russians in Korea, the

NVAF in Vietnam and the Iraqis in the Persian Gulf – time is surely running out for us.

The Vietnam War saw the maturation of many new weapons of war and we learned how to use them. Air-to-air missiles were first used en mass and were found wanting in many respects. Ground-to-air missiles affected the tactics of the war dramatically by initially forcing us to fly low into the light flak envelope. Most of our losses were to flak. ECM restored the situation and lowered losses accordingly by permitting our fighters to again penetrate at higher altitudes and out of the light flak envelope. Smart bombs were introduced and eight sorties disrupted five important bridges that had been impervious to hundreds of sorties three-five years before. Guns were re-installed in fighters and the nonsense of aircraft being too fast to dogfight was refuted again. It was again proven that we must train the way we intend to fight. And we must accept peacetime risks, costs, and perhaps even some peacetime losses to prevent more severe losses in combat.

Vietnam was a war of mistakes and clearly demonstrated how not to use fighters. Let us remember and avoid those mistakes in the future. And remember two things for sure:

• Do not start a war without a clearly stated, attainable objective.

• Recognize the limits of interdiction in a guerilla war.

NOTES

1. Green and Swanborough, *The Complete Book Of Fighters*, Smithmark Publishers, New York, 1994, pp. 454, 367, 119, 348, 500-501, 120.
2. Op. Cit.; pp. 269, 588-589.
3. Op. Cit.; pp. 150-162.
4. Op. Cit.; pp. 49-50.
5. Op. Cit.; pp. 392-398.
6. Op. Cit.; pp. 553-554.
7. Op. Cit.; pp. 396-398.
8. Eugene M. Emme, *The Impact Of Air Power*, D. Van Nostrand Company, New York, 1959, p. 202.
9. H.H. Hurt, Aerodynamics For Aviation Personnel, A Revision of NAWEPS 00-080T-80, revised by the Aerospace Safety Division, University Of Southern California, Los Angeles, CA., 1964, pp. 129.
10. Ibid.
11. Op. Cit.; pp. 201-202.
12. The author checked out in the F-100 at Nellis AFB in 1958. Then flew it in both the Philippines and Germany for a total of about 450 hours. I was also an instructor pilot in the bird as well as a maintenance test pilot.
13. Ibid.
14. The author watched several accidents by "tigers" who tried to fly F-100s like F-86s. None of them are here to discuss landing techniques. Flown with respect and flat power-on approaches, the airplane was relatively easy to land.
15. Rene Francillon, *Vietnam The War In The Air*, Arch Cape Press, New York, 1987, p. 244.
16. Enzo Angelucci, *The American Fighter*, Orion Books, New York, 1985, pp. 306-309.
17. Op. Cit.; pp. 99-102.
18. Op. Cit.; pp. 105-107, 408-409.
19. Op. Cit.; pp. 283-287.
20. Op. Cit.; pp. 105-107.
21. Francillon, Op. Cit.; pp. 130-133.
22. Ibid.
23. Mike Spick, *Fighter Pilot Tactics*, Stein and Day, New   York, 1983, pp. 136-137.
24. Lon O. Nordeen, *Air Warfare In The Missile Age*, Smithsonian Institution Press, Washington, 1985, p. 46.
25. Ibid.
26. Ibid.
27. Nordeen, Ibid.
28. Eugene M. Emme, *The Impact Of Air Power*, D. Van Nostrand, Princeton, NJ., 1959, p. 226.
29. J.E. Johnson, *The Story Of Air Fighting*, Bantam Books, New York, 1985, pp. 249-250.
30. Ibid.
31. Phillip R. Davidson, *Vietnam At War*, Presidio Press, Navato, CA., 1988, p. 260.
32. Dewey Waddell & Norm Wood, *Air War-Vietnam*, Arno Press, New York, 1978, p. 1.
33. Nordeen, Op. Cit.; pp. 9-11.
34. Ibid.
35. Ibid.
36. Ibid.
37. Waddell, Op. Cit.; pp. vii-xi
38. The author taught Air Force ROTC for four years from 1960-1964. During that time, I showed a USAAF/USAF composite strafing film in which all of the best strafing film of WW II was combined into a 50 minute feature film. I had four classes a semester, two semesters a year, and taught for four years. Counting previews, I saw that film 35-40 times. In it the Air Force Narrator flatly states that strafing was responsible for 50% of all casualties caused by aircraft in WW II. In later years, I used this statistic in arguments against Systems Command missile advocates who promised 80-90% Pks and got rid of the gun. Even at the squadron level in Systems Command we could get no listeners. Even after my operational squadron conducted actual tests with AIM-4 missiles against drones at Eglin AFB, Florida and the tests were lousy, no one would listen when we suggested that the missile's Pks were exaggerated. It was not a surprise to us when Colonel Robin Olds, commander of the Eighth Tac Fighter Wing, ordered the

AIM-4s taken off of his wing's aircraft in combat.

39. Richard P. Hallion, *Storm Over Iraq*, Smithsonian Press, Washington, DC., 1992, p. 284.
40. "Graduated Thunder" was the name Mark Clodfelter gave to Rolling Thunder in his book, *The Limits Of Air Power*. The jocks in "Nam" were less kind, we called it "Unmitigated Blunder".
41. Mark Clodfelter, *The Limits Of Air Power*, The Free Press, New York, 1989, pp. 44-46.
42. Ibid.
43. Op. Cit.; pp.42-43.
44. Op. Cit.; p. 60.
45. Nordeen, Op. Cit.; pp. 11-50.
46. Ibid.
47. Clodfelter, Op. Cit.; p. 31.
48. Nordeen, Ibid.
49. Ibid.
50. Ibid.
51. Ibid.
52. Ibid.
53. Ibid.
54. Ibid.
55. Ibid.
56. Ibid.
57. Ibid.
58. Ibid.
59. One road junction, just south of Mu Ghia pass in the Steel Tiger area of Laos, cost the author's wing nine F-4s in February of 1969. These guns were rolled in and out of caves on tracks, or rails, and cleverly camouflaged.
60. Clodfelter, Op. Cit.; p. 119.
61. Clodfelter, Op. Cit.; pp. 134-135.
62. Nordeen, Ibid.
63. Ibid.
64. Nordeen, Op. Cit.; p. 30.
65. Nordeen, Op. Cit; pp. 11-50.
66. Ibid.
67. Ibid.
68. Ibid.
69. Nordeen, Op. Cit.; p. 59.
70. Op. Cit.; pp. 68-69.
71. Op. Cit.; pp. 55-68.
72. Clodfelter, Op. Cit.; pp. 158-159.
73. Nordeen, Ibid.
74. Op. Cit.; p. 69.
75. Francillon, Op. Cit.: Appendix A, Table 2B, p. 209.
76. Clodfelter, Op. Cit.; p. 204.

CHAPTER 8

THE ARAB-ISRAELI WARS

INTRODUCTION

No study of fighters could be complete without studying the air wars between Israel and her neighbors. More than any country in the world, Israel owes its existence to the fighter and fighter pilots. Since 1948, Israel has had to fight its own "Battle Of Britain" continuously, to stay free. No other nation has so prominently displayed the results of gaining and maintaining air superiority and from the first day they became an independent nation in 1948, they have used the fighter as their primary instrument to gain this superiority/supremacy.

Another facet of the Israeli Air Force's (IAF) existence is its never ending quest to gain and maintain technological superiority, particularly in fighter weapons systems. This has been a necessity to offset the opposing Arab forces which in the better part of five decades has nearly always outnumbered the Israelis. The USSR had supplied most of this equipment to Egypt, Syria, and Iraq until the revolution in Russia shut down this supply. The Saudi Arabians and Jordan bought most of their equipment from the west, particularly, the U.S., Britain, and France, but through the last 50 years the Saudis and Jordanians have opposed the Israelis as stoutly as the rest of the Arabs.

1948--THE WAR FOR INDEPENDENCE

The IAF pre-dated the independence of the state of Israel by a number of months when the Sher Avit (Air Service) was established as part of the Haganah in the Jewish Settlement in the British mandate of Palestine on November 10, 1947.[1] On May 14, 1948, the State of Israel declared its independence. By the end of May the Zahal (Israeli Defence Force -IDF) was formed with the IAF as its air component.[2]

On the day after the Israeli declaration, the regular armies of four neighboring countries invaded Israel. Troops from Egypt, Lebanon, Syria, and Jordan partici-

---

Below: *The Israeli used any aircraft they could obtain. They fully grasped the validity of airpower. During the 1956 war, Israeli fighter pilots cut Egyptian phone lines with the **props** on their P-51s. A USAF P-51 is pictured.*

pated – plus contingents from Iraq, the Sudan, Saudi Arabia, and Morocco. The Egyptian Air Forces bombed Tel Aviv.[3] A lesser country would have been overrun in a short period of time, but due to the skill, tenacity, and courage of the IDF and the IAF, it did not happen.

Prior to this time the Israelis had negotiated the purchase of 10 Messerschmitt Me-109s from Communist Czechoslovakia and Ezer Weizman (a future commander of the IAF) and 10 other Israeli pilots with WW II fighter experience were in Prague undergoing checkout training on the Me-109s. When word reached them that Tel Aviv had been bombed, they stopped training and endeavored to get back to Israel as soon as possible. With the short legs of the Me-109, this was a chore, but they managed. The IAF procured a C-54 cargo plane and flew the first Messerschmitt into Israel in its belly. It arrived on May 20, 1948.[4]

By May 29, the first four Me-109s flew their first mission and stopped an Egyptian armored column only a few miles from Tel Aviv. It was impossible to miss the irony of Messerschmitts saving Israel. These aircraft, that during WW II were emblematic of the Nazi regime which had fathered the Jewish Holocaust, sold by Communists who rearmed the Arabs, flown by Jewish aircrews that two years earlier were flying Spitfires, literally saved Israel during these dark days.[5]

On June 3, the Egyptians tried to bomb Tel Aviv again using two C-47 transport aircraft. Both were shot down by the single operational Me-109. Meanwhile, the IAF was using every means to procure other aircraft from any available source. From this time on, the pendulum swung in favor of the IAF.[6]

By September the IAF had purchased 147 aircraft of 25 different varieties. Of the 147, 121 arrived, 33 of which were lost in combat and accidents. Only 45 aircraft were operational.[7] By now, the IAF consisted of Me-109s (six of the originals), Spitfires, Mustangs, Harvards (T-6s), Beaufighters, and B-17s.[8]

How the Beaufighter aircraft were procured is typical of the clandestine operations to procure aircraft during this period when all of the western nations were enforcing embargoes. The Beaufighters were to participate in a film about WW II in England. Elaborate sets were built and a director and camera crew hired. When the shots were scheduled, the Beaufighters equipped with long-range tanks just kept going. They were over the "Med" enroute to Israel before anyone was the wiser.[9]

There were three cease fires negotiated during the time from May 1948 to January 1949 and all combatants tried to increase combat effectiveness during the breaks with the Israelis being the most effective. Egypt and Syria (the two strongest Arab air forces) had conceded air superiority to the

Below: *One of the first fighters of the Israeli Air Force ironically was the Messerschmitt Me-109. The key fighter of the Nazi Regime protected the embryo Jewish State. The Israelis bought their Me-109s from Communist Czechoslovakia.*

Israelis by the latter part of the war.[10] Perhaps the best proof of this was the loss of three British Spitfires and one Tempest to the Israelis on 7 January 1949. It was an hour and a half before the final cease fire. The British were flying in support of the Egyptians.[11] The next day a telegram was sent to RAF headquarters in Cyprus: "Our apologies friends, but you were on the wrong side of the fence". Two British pilots were killed.[12] In aerial combat during the War For Independence, the embryonic IAF destroyed 15 Egyptian and two Syrian planes. Most of these kills were made by WW II experienced foreign volunteers hired to fly for Israel. These soldiers of fortune served Israel well. During the War Of Independence, these foreign volunteer fliers and fighter pilots were called "Machal", or "Gachal". When the war wound down, there were some troubles with these foreign pilots about salaries and execution of orders. For many of these volunteers, an oath of allegiance to Israel was a problem (they did not want to lose their American or British citizenship).[13]

In light of these and other organizational problems, Prime Minister David Ben-Gurion asked Colonel Cecil Margo, a Jewish attorney, how the IAF should be organized. Colonel Margo had been a wing commander in the South African Air Force during WW II. He conducted a detailed study and made some aggressive and forthright recommendations. Fortunately for Israel, Ben-Gurion bought them all. Despite divergent differences of opinion, the IAF became a separate service under the Air Ministry/Defense Ministry, volunteer foreign pilots were phased out,

and an extensive radar system was begun to provide both early-warning, air defense, and defensive and offensive operational control of the IAF. The technological excellence of the IAF had begun.[14]

BETWEEN THE WAR FOR INDEPENDENCE AND THE SINAI CAMPAIGN (1949-1956)

After the war ended, most of the Machal/Gachal departed Israel and the IAF got about the business of building a national, professional air force. A formal flight school was established and the first graduates matriculated in 1950. That same year, a technical school was established to produce the technicians required to equip and maintain such a force. A credo of tactical doctrine was developed and followed:

• Protection of Israeli skies.

• Achievement of air superiority by destruction of enemy air power.

• Participation in the ground war, transport, and casualty evacuation.[15]

It must be noted that no Israeli leader thought any cease fire was anything but temporary. They never knew when the "timeout" would end. This transmitted a sense of urgency to every step the Israelis took.[16]

With the economic straits of Israel, only prop aircraft were purchased during this timeframe. The decision was made to standardize on the British Mosquito and the American F-51 Mustang. The Israelis took delivery of their last Mosquitoes and Mustangs in 1952. The Arabs proceeded on into jets (the Egyptians procured British Meteors and Vampires). The Israelis subsequently bought some British Meteors in 1953.[17]

In 1955, an Egyptian/Czechoslovakian deal changed everything. The Czechs sent the Egyptians 200 Soviet jet fighters and bombers, 530 armored vehicles, 500 artillery pieces, and some naval vessels. The balance of power began to swing back to the Arabs.[18]

This turn of events forced the Israelis to make some moves. They exploited new political ties with France and bought Ouragans in 1955, and in 1956, the swept-wing Mystere IV. Because of the imbalance of forces the Israelis maintained all of their forces, both prop and jet, at readiness.[19]

THE SINAI CAMPAIGN -- 1956

The Sinai Campaign was fought as Allies with the British and French after Egypt seized the Suez Canal. For the first time the IAF conducted jet operations. The IAF com-

Below: *The Mig-17 was used by most of the Arab Countries fighting Israel.*

mitted 50 jet fighters and 50 prop fighters to operations. To begin the operation, eight pairs of Mustangs flew into the Sinai to knock out Egyptian wire communications (pole-mounted telephone wires) dragging tow cables behind the aircraft. The tow cables severed from the aircraft when they descended to pole heights and the tow cables hit the ground. The Israeli fighter pilots then went to "Plan B". They flew through the phone wires using their props to cut the wires. The missions were all successful. No F-51s were lost.[20]

The highlight of this campaign was the air drop of parachutists at Mitla Pass by the IAF. The C-47s flew at low altitude to avoid radar detection and were escorted by Ouragan

jet fighters. The 395 Israeli paratroopers took the pass and the operation was later duplicated at At-Tur. By the end of the five-day campaign, Israeli Air Forces had again seized control of the air everywhere. The IAF shot down seven Egyptian fighters and damaged two others before the cease fire.[21] This cease fire was brought about by some heavy-duty threats from the U.S. that forced Britain, France, and Israel to back off and accept Egyptian ownership of the canal.

BETWEEN THE SINAI CAMPAIGN AND THE SIX-DAY WAR

After the Sinai Campaign, the Israeli General Staff acknowledged the IAF position that air superiority was the key to battle. This was important as the resources for defense were apportioned accordingly. The IAF was given priority for defense dollars. Piston engine aircraft were retired being replaced with French Mysteres, Super Mysteres, and Vautours. In 1962 the Israelis purchased mach two Mirage IIIC fighters from France. Meanwhile the Arab nations were being re-equipped with Mig-17s, -19s, -21s, Sukhoi-7s, and Tu-16 medium bombers by the Soviet Union.[22]

After the Sinai Campaign, there followed a period of relative quiet. Egypt was busy with a war in Yemen. UN forces were in the Sinai to keep the peace between Egypt and Israel and to keep access to the port of Elath open. The IAF used this time to train and grow accustomed to their new aircraft. However, some incidents continued.

The biggest contention occurred over water. Israel began a National Water Carrier project to develop their country using water from the Jordan in the desert.[23]

In 1964 at the Arab League meeting in Cairo, Jordan, Syria, and Lebanon agreed to divert water to disrupt Israel's carrier project.[24] Conflict was inevitable as the countries sought to disrupt the other's construction projects. In response to Syrian artillery attacks, Israel responded with retaliatory attacks with tanks, artillery, and air attacks. By early 1965, the Arab attempts to divert the Jordan ended, but border incidents continued. In 1966 the newly formed Palestine Liberation Organization (PLO) began raiding northern Israel with at least the tacit approval of Syria.[25]

On Israel's other border, President Abdul Nasser of Egypt began berating his fellow Arab leaders for not supporting Syria. Nasser stepped up military preparations in the Sinai as tensions heightened.[26]

The Israelis got a break in 1966 when an Iraqi pilot defected to Israel with a brand new Mig-21. An Israeli fighter pilot was chosen to check out in the Mig and then to visit all of the Israeli fighter units, brief the pilots, and to fly mock combat with IAF squadrons. Information about this Mig was eventually used as quid pro quo to get the U.S. to approve future orders for U.S. aircraft.[27]

In July of 1966, in response to the

PLO raids and Syrian efforts to disrupt the Jordan sources in the Golan Heights, the IAF struck heavy excavation equipment working there. There were a number of dogfights between Israeli Mysteres and Mirages and Syrian Migs. In April of 1967 these fights grew more intense with heavy Syrian artillery barrages on Israeli farms and fishermen and two Mig-19s were destroyed trying to shoot down Israeli observation aircraft. Then on April 7, 1967, the IAF shot down one third of Syria's Mig-21s in three major dogfights. The IAF destroyed six Migs without loss.[28]

After this fight, the Israelis issued a stern warning to the Syrians that if these provocations continued, Israel would destroy Syria. Afraid that the Israelis might take advantage of their comparative weakness and the lack of unity within the Arab world, the Syrians appealed to the Egyptians and Russians to help them out. Nasser responded and alleged an Israeli build-up on the Syrian border. Russia confirmed this build-up. The Soviet ambassador to Israel was asked to tour the Syrian-Israeli border with the Israeli Prime Minister to prove this claim was untrue. He refused.[29]

Egyptian President Nasser ordered the UN out of the Sinai on 17 May and closed the Straits of Tiran blocking the Israeli port of Elath from access to the Red Sea. Israel was on record to the world that closing these straits would be received as a declaration of war. Nasser then moved seven divisions of his troops into the Sinai and by May 20, 1967 had over 100,000 troops with 1,000 first-line tanks on Israel's southern border. On 26

May, Nasser told the Arab Trade Union that now was the time to destroy Israel. When Jordan joined Egypt and Syria, the odds were staggering against Israel. Arab contingents were added from far away countries such as Algeria and Kuwait. The mobilized Arab forces totalled about 250,000 troops, 2,000 tanks, and over 700 first-line aircraft on three fronts, north, east, and south. The Arab world was consumed with war hysteria.[30]

After the war, the editor of the Egyptian publication El Ahram and a close confidant of Nasser's at that time, wrote that the Egyptian President made three assumptions about his actions:

• That after the UN forces were withdrawn he would close the straits of Tiran to Israeli shipping.

• That, following this action, the Israelis would likely try to open the Straits by force and break the blockade. This would lead to war.

• That, in the event of war, the ratio of forces and the state of preparedness of his forces guaranteed Egypt military success. Nasser was convinced that, in a combination of both the military and political struggle that would ensue, he would gain the upper hand.[31]

In May of 1967, most experts agreed with Nasser that the deck appeared stacked against Israel. It appeared the strength of the Arab countries was overwhelming.

Nor was the Arab/Russian timing accidental. The U.S. was totally embroiled in Vietnam with little or no forces to spare for the Middle

East.[32] The French had finished their involvement with Algeria and were no longer disposed to support Israeli military requirements. De Gaulle was anxious to restore good relations with the Arabs and their oil. In the biggest crisis of her 19 year life, Israel stood pretty much on her own.[33]

THE LINEUP ISRAEL VS. ARAB (air-to-air) -- JUNE 1967

IAF –Mirage IIIC, SNECMA Atar 09b-3 Turbojet, 13,230 #s AB, Mach 2.1, 2 x 30 MM DEFA cannon, plus 2 X Aim-9 Sidewinder AAM.[34]

Egypt –Mig-21PF, Klimov R11F2-300 Turbojet, 8,708 #s Dry, 13,490 #s AB, Mach 2.05, 1 x 23 MM twin barrel cannon, 2 x K-13 AAM.[35]

Syria –Mig-21PF, Klimov R11F2-300 Turbojet, 8,708 #s Dry, 13,490 #s AB, Mach 2.05, 1 x 23 MM twin barrel cannon, 2 x K-13 AAM.[36]

Iraq –Mig-21PF, Klimov R11F2-300 Turbojet, 8,708 #s Dry, 13,490 #s AB, Mach 2.05, 1 x 23 MM twin barrel cannon, 2 x K-13 AAM.[37]

Jordan –Hawker Hunter, Rolls-Royce Saphire, 8,000 #s dry, No burner, Mach .94, 4 x 30 MM cannon.[38]

Right: *After the 1967 War, France under President DeGaulle cut off French aircraft and support. Israel turned to the U.S. They bought a number of McDonnell-Douglas F-4Es. not surprisingly the Israelis chose the F-4E with an internal .20mm gatling gun . F-4Es were flown from the U.S. to Israel during the Yom Kippur War in 1973 using air refueling. An F-4E boom refueling is shown.*

According to the IAF who tested the defected Mig-21, the Mig-21 was very close to the Mirage IIIC in performance. The Mirage had a slightly smaller engine and was heavier. The Israelis who flew both aircraft said the Mig was slightly faster and had a stronger engine. Visibility was better from the Mirage. The Mig systems were simpler and easier to maintain. The Mig did not handle well above 500 knots and was difficult to control at these high speeds (shades of the Mig-15).[39] The

Hunters were the equivalent of the Israeli Mysteres and were inferior to their Super Mysteres.

ORDER OF BATTLE -- JUNE 1967

IAF– 72 Mirage IIIcs (three squadrons); 18 Super Mysteres (One Squadron); 50 Mystere IV-As (two squadrons); 40 Ouragans (two squadrons); 76 Fouga Magisters (two squadrons); 25 Vautours (one squadron).[40]

EAF- 120 Mig-21s (six squadrons); 80 Mig-19s (four squadrons); 150 Mig-15s & Mig-17s (five squadrons) 30 Su-7Bs (one squadron); 30 Tu-16s bombers (two squadrons); 40 Il-28 bombers (three squadrons).[41]

SAF - 36 Mig-21s (two squadrons); 90 Mig-17s (four squadrons); six Il-28s (one squadron).[42]

RJAF- 22 Hunters (one squadron); six F-104 Starfighters, which were flown out of Jordan before the war started. Most of the Jordanese fighter pilots were in the U.S. "checking- out" in the F-104. 30 F-104s were on order in the U.S.[43]

IRAQI- 20 Mig-21s (two squadrons); 15 Mig-19s (one squadron); 20 Mig-17s (two squadrons); 33 Hawker Hunters (three squadrons); 10 Il-28 bombers (one squadron); 12 Tu-16 bombers (one squadron).[44]

The IAF was outnumbered by the Egyptian Air Force alone, with equivalent quality in aircraft. Something had to be done to even the odds. That was exactly what the IAF had in mind.

THE SIX-DAY WAR

Anticipating attack by the Arabs, the IAF struck pre-emptively on the morning of June 5, 1967. The attack achieved complete surprise and was one of the most devastating in the history of air warfare. The Israeli plan was based on the initial air strikes of the Luftwaffe on the Russians during Operation Barbarossa in 1941. The Germans destroyed the air capabilities of the USSR for 18 months in that smashing initial attack. The plan was recommended for execution to the Israeli government by the then Chief Of Staff of the Israeli Defense Forces, General Yitzhak Rabin, who later became Premier of Israel.[45] Critics of present-day Israeli negotiations with the PLO may need to review Israeli history. The Golan Heights, the subject of later negotiations, were taken by Israel in this war. Rabin understood their significance. He directed the operations that took them. (Rabin was assasinated in the 1990s for discussing possible return of the Golan to Syria).

The first wave struck at exactly 0745 Cairo time on 10 Egyptian Airfields. Ten flights of four aircraft each avoided detection by flying carefully chosen routes below radar coverage over the open sea and across barren, unpopulated desert. Unobserved until actually attacking, the surprise was complete and opposition nil. These first-wave aircraft had as priority targets the Tu-16 Badger bombers and the Mig-21 air-superiority fight-

ers. The strikes were highly effective and bombed and strafed for exactly 10 minutes.[46]

As the first wave departed their targets, the second wave arrived and repeated the performance. The 10 minutes allowed sufficient time for the fighter-bombers to expend their bombs and to make several strafing passes. After a 10 minute pause, the third wave struck and beat up these bases for over 80 minutes. One of the features of the day's attack was the quick-turn of Israeli sorties averaging 15 minutes on the ground. This was a force multiplier for the IAF and illustrated the outstanding proficiency of Israeli ground crews.[47]

Although surprise was complete, the Egyptians had taken some steps to

Above: *The most formidable Arab fighter during the 1967 Six-Day War and through the Yom Kippur War in 1973 was the Mig-21. A late model Mig-21 is shown.*

deceive possible attackers. The Israeli first wave found the Egyptian Air Force had dispersed its aircraft and there were many fake aircraft in position to fool observers or attacking forces. Because of the excellence of the Israeli intelligence and the low-altitudes of the attacks, these measures were largely ineffective.

Several flights of Mig-21s were caught while taxing out for takeoff. Eight Mig-21s managed to takeoff and shoot down two Israeli fighter-bombers. Twenty (12 Mig-21s and eight Mig-19s) additional EAF fighters got airborne from the airbase at Hurghada (which had not been struck), but either were shot down by the Mirages or ran out of fuel and had to crash land.[48]

Above: *F-4 Phantom's radar missile (shown here) was the AIM-7 Sparrow. The missile guided on CW radiation and the F-4 pilot had to keep the target illuminated with radar energy until missile impact.*

Both Jordan and Syria attacked targets in Israel in support of Egypt. About noon, the IAF turned its full attention to these air forces and their facilities. By the end of the day, the Israelis had destroyed more than 300 Arab aircraft, 25 Arab air bases had been heavily hit, Jordan's air force was all but destroyed, and Egypt and Syria were stunned by heavy losses.[49] In one blow, the IAF had unstacked the deck. The Israeli forces were successful in executing an intricate, well coordinated air attack that completely turned the tables on the Arabs.

According to after-war Israeli communiques, the IAF lost 19 aircraft on the first day: four Super-Mysteres, four Mysteres, four Ouragans, four Magisters, two Mirages, and one Vautour (Note: I take all loss reports by Israelis with a grain of salt – sometimes they are accurate, at other times they do not appear to be).[50]

By the end of six days, the Israelis had destroyed 451 Arab aircraft, 391 on the ground, and 60 in the air. Twenty-six airfields were unusable. Flying close support, the IAF destroyed some 500 tanks, many other vehicles, gun batteries, dugouts, camps, supply bases, and numerous radar stations.[51] Air supe-

riority had become air supremacy during the first two days

The Israelis lost 45 aircraft during the war. At least two thirds were lost to AAA. During the war the IAF reported only four aircraft lost by air-to-air combat, but several years after the war General Ben Peled disclosed that Israel had lost 10 in dogfights.[52] Incidentally, the Israelis found the Iraqis to be the best of the Arab pilots and the Syrians the worst. The nimble Mig-17 was the toughest target while the Mig-21 could out-accelerate and out-climb the Mirage. The Mig-21 did not take hits well and burned easily while the Mig-17 could take more 30 MM hits and still get home.[53]

Air-to-air and surface-to-air missiles played little part in the kills as did electronic warfare.[54] Most of the air-

to-air kills were by the 30 MM cannons installed in the Mirage IIICs. Halfway around the world at the same time, Americans were fighting the same Soviet aircraft much less successfully with F-4 aircraft not equipped with guns.[55]

After the war, in an after-action report, the IAF listed five reasons for their outstanding success in the Six-Day War:

• Simplicity of planning and faith in success.

• Almost total compatibility of planning and implementation.

• Precise execution of orders by pilots.

• Centralized control.

• Precise intelligence.[56]

WAR OF ATTRITION AND AFTER -- 1967-1973

The USSR sprang to the mark to replace all of the Arab losses with new and improved equipment. Egypt received 130 jet fighters by the end of June. "Stopped" by an armistice, the Six Day War devolved into a series of incidents: artillery barrages, commando raids, dogfights, reconnaissance flights, surface-to-air missile engagements, etc. It was what the Iraelis called the "War of Attrition" where Israelis launched punitive raids after incursions by the Egyptians.

After the Six Day War, with DeGaulle's embargo on the pur-

chase of French airplanes by the IAF, Israel turned to America for new aircraft. Israel also decided to try and develop her own fighters. Negotiations with America went very slowly. The first American entry on the scene was the Douglas A-4 Skyhawk. Israel had signed an agreement for Skyhawks before the Six Day War, but the war delayed delivery. The Skyhawks were finally delivered in early 1968. The Israelis then procured the American "top of the line" fighter, the McDonnell-Douglas F-4 Phantom II in 1969. It was no accident that the Israelis chose F-4Es with a 20 MM Gatling gun installed in the nose.[57]

In addition, Israel's second attempt at building her own fighters (the Nesher was first) entered the lists:

the Kfir was an upgraded Mirage with forward strakes to improve maneuverability and powered by a GE J-79 engine, two of which powered the Phantom II. The J-52 that powered the A-4 was also installed in some Mysteres.[58]

On March 8, 1969, Nasser announced his war of attrition using his numerical superiority and plentiful Soviet equipment. Despite intense Israeli retaliation for every incursion, Egyptian raids and dogfights continued until July 1969. During these fights both sides claimed kills with very few losses. The Israelis evidently came out on top however as Egypt abruptly ceased air incursions and dogfights in July.[59] Egypt had gone from the "attritor" to "attritee".

On September 7, 1969, the IAF accepted its first four F-4Es. Two days later, 60 to 70 Egyptian fighter-bombers caused extensive damage to Israeli fortifications and encampments, but at high cost, losing seven Mig-21s, three

Left: *Mirage IIIC airframes were modified to be more maneuverable and a GE J79 was installed versus its Snecma Atar. The result was a faster more maneuverable and more reliable fighter –the KFIR shown here. Note the "strakes" above and to the rear of the engine intake.*

Su-7Bs, and one Mig-17 to fighters, Hawk SAMs, and AAA. During the fall, the Egyptians with almost unlimited equipment from the USSR tried to finalize and integrate their air defense network to shut down Israelis retaliatory strikes.[60]

In January 1970, Israel changed its tactics and began to raid deep into Egypt using their new F-4s. After several successful Israeli deep raids, Nasser flew to Moscow requesting a deep strike capability to retaliate. The Russians refused but offered defensive assistance in the form of a defensive system equipped and manned by Russians. This incursion into the Mideast by a major power was accepted by Egypt and was viewed with jaundiced eyes by both Israel and the U.S.[61]

The Israelis were very leery because they were afraid of provoking an incident with the USSR which would increase their involvement in the Middle East. Nor was the U.S. enamored with that prospect. As Russian involvement accelerated, incidents were sure to occur. The Israeli monitors were picking up Russian language everywhere both in the air and on the ground. Russian pilots and SAM technicians were of particular concern.[62]

On April 30, 1970, Soviet manned Egyptian Migs moved to intercept an Israeli reconnaissance flight. When the Israelis moved to protect the flight, the Russians backed off. Concerned with a possible incident with the Soviets, the Israelis stopped the deep raids into Egypt. For three months the IAF maintained a low profile on the Egyptian Front along the Suez Cannal.[63] During July, Israeli lost a number of F-4s to missiles along the Egyptian front. The Soviets had integrated the Egyptian air defense system and the IAF was running into several SA-2s and SA-3s fired at a time. The IAF fighter pilots found the SA-3 much more difficult to outmaneuver than the SA-2. Several were hit by AAA or a second or third missiles.[64] There were portents of what was to come during this month, and for once the Israelis were not as alert as they should have been.

On July 25, Skyhawks attacked Egyptian positions along the Canal and the Russian Migs appeared. The IAF A-4s were told to disengage and when they fled, the Russians pursued across the Canal into the Sinai. One VVS fighter fired an IR missile which struck a Skyhawk but he landed safely.[65]

Several days later, the Israelis set a trap for the Soviet advisers. They decided to attempt to get a clear victory over the USSR forces to embarrass the Russians. On July 30, using eight Mirages as bait, the IAF got 24 Russian Mig-21s to scramble. Ambushing them from below, where Phantoms had been hiding in the weeds, the IAF shot down five Migs with no loss to themselves.[66]

Another cease fire was negotiated on August 7, 1970 by the U.S. Secretary of State. The IAF assumed a cease fire date of August 4. Just seconds before the Aug. 4 cease fire was to go into effect, the IAF bombed and dropped a bridge across the Suez into the canal. As the four F-4s pulled out of their diving attacks their cease fire went into effect. The War Of Attri-

tion was over for the IAF.[67]

THE YOM KIPPUR WAR

From 1970 to 1973 there were a number of changes between Israel and Egypt. The first came when Gamel Abdel Nasser died in 1970 and was replaced by Anwar Sadat as President of Egypt. There was an immediate attempt at a coup (many said inspired by the Russians).[68]

Sadat tried unsuccessfully to gain more favorable terms than the U.S. Plan that Nasser had rejected, but the U.S. no longer felt bound by this plan. A new stalemate developed. An oil crisis in the U.S. had precipitated a fuel crisis in the U.S. and the prices of gasoline had tripled. The Arabs felt they had new leverage in dealing with the U.S.[69]

Sometime during this period, Sadat decided on war as his only alternative.[70] In the intervening years of the cease fire, the Egyptians, with extensive Russian help, had continued to reinforce their air defense system. By the eve of the Yom Kippur War in October 1973 this system was built up to the greatest depth and mix of air defense systems ever seen.[71] AAA was integrated with SA-2, SA-3, SA-6, and SA-7 shoulder held missiles. The SA-6 in particular, was a new, formidable weapon for which the Israelis had no electronic

defense. The AAA included the deadly radar guided ZSU-23-4.[72]

In addition, Sadat had coordinated his plans with Syria and they had prepared for the war for over six months in secret. In Egypt, over 150 Sam sites (about 60 along the Canal) were hardened and integrated into a coherent air defense system along the Canal and in the Nile Delta. In addition, mobile SA-6 missiles were used along the Canal, at key points, were integral to army units, and had been integrated into the defense system. Scores of ZSU-23s were tied in with organic 12.7 MM, 14.5 MM, 37 MM, 57 MM, and the SAMs. Most of these units could be accurately fired with optical sights as well as radar. During the War of Attrition all of the Egyptian airfields had been hardened. Aircraft hangars were reconstructed as concrete, blast-resistant shelters. The aircraft were serviced and maintained in these shelters. Runways and taxiways were reconstructed to be super resistant to bombings. These improvements made it impossible to duplicate the massive destruction of Arab air power done during the first strikes of the Six-Day War.[73]

In Syria, these same defenses were emplaced in depth from the Golan Heights all the way to Damascus. Israeli aircraft were faced with a veritable wall of fire to the north over Syria and to the west over

Above: *The Israelis encountered the Hawker Hunter when Iraq & Jordan entered the 1967 War. The Israeli Mirages had little problem with this fine sub-sonic fighter. Iraq was still using the Hunter in 1973–during the Yom Kippur War.*

Egypt. The Arabs were much less vulnerable to air attack, had modern equipment, were well-trained, and were well-supplied by the Soviets. With these defenses, they were formidable opponents.[74]

Syrians gained air superiority over their battlefields without flying their aircraft although Arab aircraft attacked targets all over Israel. *For the first time in the history of warfare, air superiority over battlefields was gained by surface-to-air missiles.*

For once, Israeli's usually superb intelligence failed (warning was not received of a confirmed attack until just hours before H-hour). It appears the Israelis were a bit complacent and they did not believe the Arabs would dare attack them. There was significant data to prove the Arab capabilities, but the Israelis reacted on what they thought were Arab intentions. Consequently, when the Arabs struck, the Israelis were not mobilized and many reserves were not called up, until just before the attack and some not until after the shooting started.[75]

At 1400 October 6, 1973, the Egyptians and Syrians struck. There were only regular IDF forces in place, which were vastly outnumbered, and the IAF. These forces had to hold until Israeli reservists could be mobilized to reinforce them. The IAF needed to support the skeleton forces holding both fronts to repel massive armored thrusts, and had to penetrate airspace controlled by SAMs to do it. The IAF losses were horrendous (in my opinion probably higher than has been reported even now).

The Arabs achieved surprise by attacking on Yom Kippur and Ramadan (the two holidays fell on the same day) in an unexpected way, with unprecedented violence. On October 6, 1973, the Egyptians and

ISRAELI ORDER OF BATTLE

The IAF consisted of the following at the start of the war:

- 127 F-4E Phantom II fighters

- 162 A-4E, A-4H, A-4N Skyhawk fighter-bombers

- 35 Mirage IIIC fighters

- 40 Nesher fighter-bombers (Israeli built Mirage Vs)

- 15 Super Mystere fighter-bombers [76]

EGYPTIAN ORDER OF BATTLE

The EAF consisted of the following at the start of the war:

- 210 Mig-21F, Mig-21PF, Mig-21MF fighter-interceptors

- 100 Mig-17 fighter-bombers

- 80 Su-7D fighter bombers

- 5 Il-28 bombers

- 25 Tu-16 bombers [77]

SYRIAN AIR ORDER OF BATTLE

The SAF consisted of the following at the start of the war:

- 200 Mig-21 fighter interceptors

- 80 Mig-17 fighter-bombers

- 30 Su-7 fighter bombers

OTHER ARAB FORCES

Other Arab countries contributed the following:

• Iraq–one squadron of Hunters in Egypt, one squadron of Mig-21s in Syria, two squadrons of Su-7Bs in Syria

• Algeria–one squadron of Su-7Bs in Egypt

• Libya–two squadrons of Mirages

The IAF was less influential in the Yom Kippur War than in the Six-Day War. This was largely due to the integrated defense systems used by the Arabs and the better equipment and training of the Arab Air Forces. The Arab defense systems prevented effective close support until these defenses could be suppressed.[78]

The Egyptian and Syrian armored forces were initially successful and drove the undermanned IDF back everywhere. The IAF was relatively ineffective even though they attacked ferociously with some success in spite of their heavy losses. On the Golan, the IAF lost 30 aircraft the first day. That is almost 8% of the IAF total combat force and the other half of the IAF was sustaining similar losses against the Egyptians. Losses were so high during the first few days that ground forces observing the "shootdowns" began to refuse to request air support.[79]

Destruction of the Arab air forces on the ground was no longer possible with IAF weapons as the hardened shelters had made air attacks against Arab airfields impractical. The Israeli forces were fairly well-equipped with American Vietnam-era ECM gear which provided some protection from SA-2 and SA-3 systems, but there was no detection or counter measure capability against the SA-6 missile systems. The Arabs used new tactics firing missiles in clusters and the IAF aircraft would avoid one or two missiles only to be hit by the third or fourth missile. The IAF had no choice but to go in low and take their chances. As in Vietnam, they ran into heavy barrage fire around defended targets. Over 84% of IAF losses were to AAA, particularly, light automatic weapons. In addition the SA-6s and shoulder mounted SA-7s were also effective at low altitude. For the first time in IAF history the air reduction of defended targets was not the most cost effective method of attack. In fact on the Egyptian front, ground forces were used to remove the air defense systems before the IAF resumed attacks.[80]

Although exact numbers cannot be determined, Israeli proficiency in air-to-air combat appears to have continued. According to Israeli sources, the Arabs lost a total of 451 aircraft to all causes. Most of these fell to Israeli fighters. Israeli fighters accounted for at least 370 Arab fighters, one Tu-16 bomber, and 40 Helicopters, while losing only four Israeli fighters.[81] The same sources stated that the Israelis lost 115 aircraft, 10 by accident, one by friendly fire, 48 by SAMs, 52 by

AAA, and the four by air combat.[82]

Some IAF claims are unbelievable. Four IAF losses air to air seems a bit tainted. One Egyptian Mig-21 regiment claimed 22 kills in air combat and displayed gun camera film that supported a number of the kills. The EAF maintained the IAF lost 303 fighters and 25 helicopters on the Sinai front alone. The EAF commander also maintained that the Israelis had 550 aircraft at the start of the war well over 100 more than the IAF quoted at the beginning of the war.[83]

According to U.S. intelligence, Arab AAA and SAMs cost the Israelis 80% of their losses, while 10% were lost air to air. The same U.S. sources list Arab losses as: 242 Egyptian, 179 Syrian, and 21 Iraqi to all causes. This was within

nine of the Israeli claims.[84]   Of interest also was the IAF's proclivity to use the gun rather than air-to-air missiles. This was often dictated by the large-scale swirling melees characteristic of this war.[85]   One can only wonder what would have happened to U.S. F-4D or F-4J aircraft in these types of fights with no guns. Or, to an F-22 today, if it were bought without a gun, as has been discussed.

AIRLIFT REPLENISHMENT

Both sides received substantial airlift from their super power sponsors. Soviet An-12 and An-22 transports flew more than 900 round trips starting from day one (the Soviets were obviously very aware of the war's start date). Starting much later in the war,

the U.S. flew 566 round trips, flew much longer distances from the U.S., to the Azores, and down the Mediterranean without landing (none of our so-called allies – Spain and France in particular would permit U.S. reinforcements to land). U.S. tonnage exceeded that delivered by the USSR. In addition, 48 F-4Es and 80 A-4 Skyhawks, were delivered, most of which flew in themselves along the same torturous route. This replenishment included Sidewinder AAMs, Shrike anti-radar missiles, Walleye glide bombs, Maverick air-to-ground missiles, and TOW anti-tank missiles.[86]   In the author's opinion, U.S. airlift saved Israel from defeat. Prime Minister Golda Meir said: "For generations to come, all will be told of the miracle of the immense planes from the United States bringing in material that meant life to our people"[87].

Left: *Many of the Arab Nations were supplied with Su-7 ground attack aircraft by the Soviet Union. Shown is the two-seater trainer version. The Su-17 was a swing-wing version of this ground attack aircraft.*

DEFCON 3

It is not common knowledge, but the USSR became very alarmed when the IDF crossed the Suez and drove in proximity of Cairo during the Yom Kippur War. They made preparations to heavily reinforce Egypt with Soviet forces. President Nixon, caught in the throes of Watergate, raised U.S. defense status to defense condition three (DEFCON 3), the highest peacetime alert status for U.S. forces. He then bluntly told the USSR to butt out of the Middle East. In spite of Democratic Party protests that he was trying to divert attention from Watergate, he adamantly held his ground and the Russians finally backed down.[88] This confrontation is not well known, but we were again at the brink of nuclear war and Nixon once again demonstrated his moral courage in the international arena. I believe this decision will go down in history as one of our greatest.

ENTEBBE

Although not a fighter operation, the successful Israeli rescue of hostages taken to Uganda in 1976 is a classic in all of history, particularly in light of the American hostage fiasco in Iran several years later. Of particular interest is the fact that though the rescue was planned at the highest levels of the IDF and approved by the Israeli Cabinet, command was given to one man, an Israeli Lieutenant Colonel. It was a small unit action so Israel gave the job to their

best small unit commander and gave him the authority and responsibility needed for the job.[89] This delegation of authority and unity of command appears to have been the major difference from the U.S. abortive rescue attempt into Teheran.

The Israelis know how to delegate authority to commanders in the field. They then support them in every way possible, and get out of their way and let them do their job. It is a formula for success that many U.S. officials need to learn. Ironically, the unit commander was the only casualty at Entebbe. He was killed by a sniper's bullet while directing the hostage evacuation.

THE OSIRAK REACTOR -- 1981

On June 7, 1981 at 1600 Tel Aviv

spent over the target was no more than two minutes. The attack was made with "dumb" bombs using the F-16 automatic bombing system, an improvement on the Nav-Weapons System used in the A-4H and A-4N in the 1970s (first used in the USAF A-7D and later the USN A-7E). The official Israeli communique summed it up: "The Israeli Air Force yesterday attacked and destroyed completely the Osirak nuclear reactor which is near Baghdad. All Of our planes returned home safely".[91]

One might wonder about Desert Storm 10 years later if that mission had not have been so successful. If Iraqi "Scud" missiles had been equipped with nuclear warheads, the Persian Gulf might have been Armageddon.

LEBANON/BEKAA VALLEY -- 1982

The first major action by the IAF after the Yom Kippur War was "Operation Peace for Galilee". When the Palestine Liberation Organization was ejected from Jordan in 1970, they took refuge in Lebanon. Lebanon was torn by factions (the Syrian Army, the Druse, and the Christian Phalangists Militia) and had a UN demilitarized zone. Seven hundred PLO guerillas set up operations in the UN zone and this group began terrorist operations against

time, eight F-16s and six F-15s heavily laden with fuel and ordnance lifted off from Etzion air base in the Sinai. There was nothing unusual about the take-offs as similar Israeli training missions were flown from Etzion everyday. But as these fighters joined in formation, they turned to the east.[90]

They flew a diverse route over barren desert at 30 meters. More than an hour later, the two flights arrive at their target: the Osirak Reactor at Tuwaitha, 17 kilometers from Baghdad. The F-16s popped up to altitude and delivered their bombs exactly on target. The reactor collapsed on itself and the subsequent chain reaction destroyed the nuclear core without spreading dangerous radiation all over the surrounding countryside. The time the Jews in Galilee. This terror steadily increased until Israel had enough. Heavy shelling of Galilee with multiple launcher rocket systems (MLRS) and the PLO murder of the Israeli ambassador to Britain were the straws that broke the camel's back and caused Israel to attack the PLO in Lebanon.

The official statement outlined the objectives on June 5, 1981 as the Israeli armored forces rolled into Lebanon:[92]

• Order the IDF to place the civilian population of the Galilee beyond the range of terrorists' fire from Lebanon, where they, their bases, and their headquarters were concentrated.

Below: *The F-16 shown here and the F-15 were highly effective weapons systems in the attack on the Osirak Reactor and in the Bekaa Valley. In the Bekaa the AIM-9L missile shown (top-outboard) was decisive air to air, and the Harm missile shown (inboard) were decisive in destroying the Syrian integrated air defense system.*

• Name the operation "Peace For Galilee".

• Order that the Syrian Army not be attacked unless it attacks our forces.

• Aspire to the signing of a peace treaty with independent Lebanon, its territorial integrity preserved.

When the Syrian Air Force decided to intervene, the IAF was ready after eight years of study of the Yom Kippur War and the integration of sophisticated new equipment . Its primary fighters were the new American F-15 and F-16. New, more effective, American ECM, AAMs, SAMs, and AGMs were operational and most had been improved by Israeli engineers.[93] The IAF was also still using the A-4s and F-4s of the Yom Kippur War and had introduced the Kfir we discussed earlier. The F-15, F-16, and Kfir were used for air to air and the F-4, A-4, for air to ground. The F-16 and Kfir were sometimes used in either role. The Iraelis also bought E-2

Hawkeyes from the U.S. to control their air operations.[94]

The IAF had used the shock of the heavy losses during the Yom Kippur War as a springboard to operational

154

improvement and research and development into how to overcome concentrated missile defenses.[95]

During the 1970s the Syrian Air force was significantly upgraded and replaced with new Soviet equipment. They received upgraded Mig-21s, and new Mig-23, Mig-25, and Su-20 fighters. The Syrians also expanded pilot training to develop better pilots and more of them.[96]

Egypt and Israel concluded a peace treaty in 1979 after continued years of strife. Skirmishing continued with the PLO and Syria. In the air, dogfighting between Syria and Israel was common. These skirmishes intensified after Israel and Egypt made peace. The F-15 proceeded to demonstrate its superiority by

destroying every brand of Mig in these dogfights with no losses. American AAMs, as modified by the Israelis, also demonstrated a clear superiority. On March 13, 1981, the best of the west met the best of the east when a Mig-25 Foxbat attacking an IAF recon aircraft was destroyed by an Israeli F-15 launched Sparrow.[97]

The Syrians in Lebanon based their air defense in the Bekaa Valley on their experience in the Yom Kippur War with belts of missiles, integrated with radar controlled AAA and optically capable guns and missiles. The Soviets had added SA-8 and SA-9 missiles to upgrade this inventory.[98]

The third day after the Israeli tanks rolled into Lebanon, the SAF sortied in support of Syrian troops with improved Mig-21s and Mig-23s. In a

Above: *The Mig-23 could not survive in airspace full of hostile Israeli F-16s & F-15s.*

series of dog fights, six Migs were shot down, most, by F-15 missiles. No Israeli aircraft were lost. Israeli aircraft were using unprecedented ECM. Syrian fighter communications and radars were jammed as they lifted off on takeoff. Israeli fighter attacks were often a total surprise as the Syrians were not always in contact with their radars and operational centers and controllers.[99]

When the IDF decided to attack Syrian forces in Lebanon, the first order of business was to take out the missile forces in the Bekaa Valley. The IAF reconnaissance had pinpointed the location of most of the missile batteries and their attack was a massive, coordinated strike that had been planned for months. It included barrage jammers, state-of-the-art jamming pods on each fighter, and an entirely new array of missiles and drones and smart bombs. The strike force was configured with twice the amount of ordnance needed to obliterate the targets.[100]

To the Syrians, the strike was a massive, simultaneous strike, but unbeknownst to them the first waves were drones that simulat-

---

Below: *The Mig-25 was designed to intercept Mach 3 high altitude bombers. It is a heavy, very fast fighter (perhaps the fastest in the world) with a powerful radar, and look-down, shoot-down capability. It has proven no match for F-15s in maneuvering fights.*

---

ed aircraft; some "dumb", some "smart". When the Syrian defense forces responded, Israeli aircraft fired radar homing missiles (Israel modified Shrikes, Standard Arms, and Zeevs [a ground-to-ground missile with the last stage an Israeli Shrike] which retained the targets in memory when the target radar shutdown. The defense systems were decimated.[101]

As each target radar had been identified and ECM countermeasures built into the strike force, the remaining sensors were blinded. With the defense system down, conventional attackers equipped with "smart" weapons moved in, and mopped up. The system was obliterated and never recovered.[102] An integrated system of 23 missile batteries was wiped out with *no aircraft losses!*[103] –a far cry from the Yom Kippur War. The IAF simultaneously flew hundreds of interdiction sorties destroying the vanguard of the

Syrian Third Armored division.[104]

AIR-TO-AIR

The predominance of the IAF in air-to-air operations was staggering. The SAF responded to IAF incursions with waves of Mig-21s, Su-22s, Mig-23s, and some Mig-25s. F-16s, in the Israeli inventory only two years, shot down 44 Migs. Forty were accounted for by F-15s, and one by an F-4. During the years since the F-15 was introduced, until the end of the Galilee Operation, the F-15 ran up a 58-0 score over the Arabs. I say again, 58 to zip! According to U.S. sources, only seven percent of the Israeli kills were with the gun. Missile effectiveness was at an all-time high. The Israeli Shafir and Python did well. The American Aim-9L got very high marks, as did the Sparrow.[105]

Israeli recon, command and control, and drone operations were superb and the IAF was able to disrupt Syrian command, control, and communications.[106]

SUMMARY

Israeli air operations have been based on the fighter and gaining air superiority. Their success is well known and much of that success can be attributed to the proper use of fighter weapons systems. There are probably no better examples of effective and innovative fighter operations.

One of the earliest examples was the improvisation in cutting Arab telephone lines with fighter propellers during the 1956 war. The incredible destruction of the Arab air forces during the Six-Day War was another.

But the Israeli Wars clearly demonstrate the postulates and corrollaries of Air Power determined in other wars in other places. One small fight that the Israelis lost in the 1967 Six-Day War, confirmed our experiences in Vietnam in 1965. On June 6, a flight of Mirages trying to penetrate to a deep target in Iraq were caught at low altitude by Iraqi Hunters. The IAF lost three supersonic Mirages and the Iraqis, a Hunter.[107]

Supersonic aircraft caught at low altitude were once again easier targets than one would expect for subsonic aircraft that could outmaneuver them. The more powerful Mach two aircraft, laden with bombs and other ordnance, were unable to use their power to maneuver vertically. As we have noted before, a supersonic fighter-bomber loaded down with external stores is just a subsonic bomber, whether in Vietnam or over Iraq. Hunters and Mig-17s were effective if they could attack Mach two fighter bombers on the deck with little or no warning.

Unlike all the other Arab-Israeli wars and conflicts, the Yom Kippur War was a bad experience for the IAF. They lost heavily in the early days of the war after being caught unready on Yom Kippur.

The IAF threw itself at the Egyptians and Syrians with a ferocity born of desperation. The integrated defense systems of the Arabs fought the IAF to a virtual standstill.

During the Yom Kippur War, the IAF flew an average of more than 500 sorties/day for 18 days for a total of more than 10,000 sorties. If Israeli numbers are to be believed, the IAF lost one aircraft per 100 sorties, while the lost rate during the Six-Day War was four for each 100 sorties. No exact sortie figures are available for the Arabs, but best estimates are between 9,000 and 10,000.[108]

The Soviet air defense systems used by the Arabs are the most discussed features of the Yom

Kippur War. They provided integrated defenses from ground level to over 60,000 feet. The SA-2s, SA-3s, SA-6s, and SA-7s interlocked with massive AAA proved formidable in the defense. The SA-6 and ZSU-23 were particularly effective. Soviet missiles were fired in clusters and there was no really effective counter measure to this method of defensive attack.

The Arab Air Forces capability to fight air-to-air clearly was not as good as the Israeli systems. As we saw in Korea, Migs often maneuvered to a position of advantage and were unable to obtain a kill. Examination of the Soviet AAMs showed G limitations similar to the early Sidewinders, as well as a gyroscopic gunsight that did not work when the aircraft were banked and loaded-up above 2.75 Gs.[109]

The Arabs flew "finger four" formation similar to many major air forces. The IAF used a variation similar to the U.S. Navy's "loose deuce".[110]   Normally the difference in the two is in the finger four, the leader usually attacks while the wingman protects him. In the loose deuce, either the leader or the wingman can attack while supporting and coordinating their tactics. Most fighter pilots feel the loose deuce is a refinement/improvement on the finger four.

It was clear in the Yom Kippur War that the day of the AA missile was arriving (late model Sidewinder and Shafir IR missiles had a 56% kill rate [Pk] and over 200 kills, while the Sparrow, used only by F-4s and sparingly, got seven kills for 17 launches for a 42% Pk). The SAMs used by both sides changed the entire complexion of the war.[111]

The lesson of the Yom Kippur War in 1973 was ECM is essential in modern war. Fighter weapons systems depend on electronic capabilities to counter missile systems. Stealth anyone?
In the Bekaa Valley 10 years later, the requirement for state-of-the-art ECM was proven for all to see. The Israelis destroyed a Syrian integrated air defense network using electronic warfare, anti-radiation missiles and remotely piloted vehicles. In sharp contrast with Yom Kip-

Right: *The McDonnell-Douglas F-15 is the top operational all weather fighter in the world today. Many new fighters are entering service which will challenge this ascendancy. For the last 15–20 years the Israelis have exploited this superiority. Note the refueling boom to the right of the F-15.*

pur, Israeli losses in the Bekaa were minimal.

The nation of Israel exists because of air superiority for the years of its existence. Its leaders grasped from day one the impor- tance of the fighter in achieving air superiority and Israel's bril- liant history is really a modern history of fighter aviation. Amid nuclear proliferation throughout the world, will the importance of fighters continue? Perhaps, but one thing is certain: the reverence Israel reserves for her fighter pilots is not misplaced.

NOTES

1. Stanley M. Ulanoff & David Eshel, *The Fighting Israeli Air Force*, Arco Publishing, Inc., New York, 1985, pp. 5-8
2. Ibid.
3. Ibid.
4. Eliezer Cohen, *Israel's Best Defense, The First Full Story Of The Israeli Air Force*, Orion Books, New York, 1993, pp. 25-38.
5. Ulanoff, Ibid.
6. Ibid.
7. Cohen; Op. Cit., p. 49.
8. Ulanoff, Ibid.
9. Cohen, Op. Cit.; p. 51.
10. Ibid.
11. Ibid.
12. Op. Cit.; p. 59.
13. Op. Cit.; pp. 38-42.
14. Ibid.
15. Ulanoff, Op. Cit.; p. 8.
16. Ibid.
17. Ibid.
18. Ibid.
19. Ibid.
20. Ibid.
21. Op. Cit.; p. 10.
22. Op. Cit.; p. 9.
23. Ibid.
24. Lon O. Nordeen Jr., *Air Warfare In the Missile Age*, Smithsonian Institution, Washington, D.C., 1985, pp. 111-123.
25. Ibid.
26. Ibid.
27. Cohen, Op. Cit.; pp. 186-189.
28. Op.Cit.; p. 190.
29. Chaim Herzog, *The Arab-Israeli Wars*, Random House, New York, 1982, pp. 157-163.
30. Ibid.
31. Ibid.
32. In 1967, the author was a tactical evaluation officer at 17th Air Force Headquarters in Europe. In my job, it was necessary to study all the contingency plans in USAFE. In my opinion, our commitment in Vietnam would have made USAFE plans for a Middle East crisis very difficult to execute. In the event of an Israeli defeat by the Arabs, those plans might have been impossible without a major war escalation or mobilization.
33. Herzog, Ibid.
34. Green & Swanborough, *The Complete Book Of Fighters*, Smithmark, New York, 1994, pp. 151-152.
35. Op. Cit.; pp. 396-397.
36. Ibid.
37. Ibid.
38. Op. Cit.; pp. 291-292.
39. Cohen, Op. Cit.; p. 189.
40. Nordeen, Op. Cit.; p. 112.
41. Op.Cit.; p. 114.
42. Op. Cit.; p. 115.
43. Ibid.
44. Op. Cit.; p. 116.
45. Cohen, Op. Cit.; pp. 191-192.
46. Nordeen, Op. Cit.; pp. 116-121.
47. Ibid.
48. Ibid.
49. Ibid.
50. Ibid.
51. Ulanoff, Op. Cit.; pp. 12-13.
52. Nordeen, Op. Cit.; p. 122.
53. Op. Cit.; p. 123.
54. Ibid.

55. Ibid.
56. Ulanoff, Op. Cit.; p. 13.
57. Ibid.
58. Green & Swanborough, Op. Cit.; p. 302.
59. Nordeen, Op. Cit.; p. 129.
60. Ibid.
61. Op.Cit.; pp. 133-134.
62. Cohen, Op. Cit.; p. 297.
63. Op. Cit.; pp. 305-310.
64. Op. Cit.; p. 299.
65. Ibid.
66. Op. Cit.; pp. 306-309.
67. Ibid.
68. Peter Allen, *The Yom Kippur War*, Charles Scribner & Sons, New   York, 1982, pp. 27-44.
69. Ibid.
70. Ibid.
71. Ulanoff, Op. Cit.; pp. 74-89.
72. Nordeen, Op. Cit.; pp. 143-172.
73. Ibid.
74. Ibid.
75. Cohen, Op. Cit.; p. 323.
76. Nordeen, Op. Cit.; p. 147.
77. Op. Cit.; p. 148.
78. Op. Cit.; p. 162.
79. Ulanoff, Op. Cit.; pp. 80-83.
80. Ibid.
81. Nordeen, Op. Cit.; p. 163.
82. Ibid.
83. Ibid.
84. Ibid.
85. Ulanoff, Op. Cit.; p. 80.
86. Nordeen, Op. Cit; p. 163.
87. Ulanoff, Op. Cit.; p. 89.
88. Allen, Op. Cit.; pp. 229-234.
89. Ulanoff, Op. Cit.; pp. 119-121.
90. Op. Cit.; pp. 129-133.
91. Ibid.
92. Ulanoff, Op. Cit.; pp. 136-152.
93. Ibid.
94. Nordeen, Op. Cit.; pp. 177-184.
95. Cohen, Op. Cit.; pp. 470-471.
96. Ibid.
97. Ibid.
98. Ibid.
99. Cohen, Op. Cit.; pp. 465-477.
100. Ibid.
101. Ibid.
102. Ibid.
103. Nordeen, Ibid.
104. Ulanoff, Op. Cit.; p. 137.
105. Ibid.
106. Op.Cit.; p. 185.
107. Nordeen, Op. Cit.; pp. 116-121.
108. Nordeen, Op. Cit.; p. 163.
109. Nordeen, Op. Cit.; p. 169.
110. Ibid.
111. Op. Cit.; p. 171.

Above: *In the Indo-Pakistani conflicts the Folland Gnat emerged a star. High power-to-weight ratios, small size and superb maneuverability. But these wars were before the advent of modern air-to-air, all aspect missiles. Gun attacks were primary.*

CHAPTER 9

THE INDIA-PAKISTAN1 CONFLICTS

THE KASHMIR CONFLICT AUGUST 1965

INTRODUCTION

Conflict between Pakistan and India is as old as conflict between Moslem and Hindu. The British recognized this in 1947 when they partitioned the country along religious lines forming Moslem Pakistan and Hindu India. This division caused real trouble about the province of Kashmir. Under a UN mandate in 1964, Kashmir was partitioned: one third to Pakistan and two thirds to India. The province was supposed to determine by referendum to whom they were to belong. The vote was never held and when India decreed that its portion of Kashmir was to become Indian, trouble erupted. In August 1965, open, undeclared warfare broke out between the two countries.[1]

The Air War portion of the conflict was of interest to military advisors around the world as it provided combat trials to many western fighters that had not been compared in anger before.

Air Order Of Battle

Before 1956 the Pakistani Air Force (PAF) used primarily British equipment, but in 1956, a U.S. Military Defense Assistance Program (MDAP) was begun. On the eve of the Kashmir Conflict, the Pakistani order of battle was:

• 100 F-86F Sabres

• 25 B-57 Canberras

• 12 F-104A & F-104B Starfighters

• 12 T-33A trainer/fighter/bombers (two seat F-80s)[2]

Twenty-five percent of the F-86s were fitted with AIM-9B sidewinders. Pakistani training was excellent, beginning in the T-6 and then progressing to the T-33.

The Indian Air Force (IAF) was larger than the PAF and had a myriad of equipment:

• 8 Mig-21 Fishbeds

• 118 Hawker Hunters

• 80 Mystere IV-As

• 50 Gnats

• 56 Ouragans

• 132 Vampires

The IAF also had a number of Soviet SA-2 SAMs installed around primary facilities.[3]

AIR BATTLES AND ANALYSIS

Lasting only three weeks, this conflict was relatively bloody. Pakistan claimed to have killed or captured 8,200 Indian soldiers, destroyed 110 aircraft, and to have destroyed or captured 500 tanks.[4]

Right: *India was equipped with USSR fighters like this Mig-17*

India, on the other hand, claimed to have killed or captured 5,259 Pakistanis, and destroyed 73 aircraft and 471 tanks.[5] Both sets of claims appear to be grossly exaggerated. India admitted the loss of only 33 aircraft, and Pakistan 19.[6]

The air combat took place at relatively low altitude. The F-86F appeared to have an ascendancy over the Hunter. The Gnat on the other hand was the surprise of the conflict. Nicknamed the "Sabre Slayer", it accounted for at least six Sabres. Its small size made it hard to see and its out-

standing power to weight ratio accounted for superb maneuverability.[7]

The Mach 2 F-104 was also successful shooting down two Mysteres and two Canberras. The F-86Fs allegedly shot down the most aircraft: 19 Hunters, six Gnats, four Vampires, one Mystere, and one Canberra. The long awaited fight between the F-104 and Mig-21 did not occur in 1965. Aim-9Bs got 9 kills for 33 launches proving again that missile brochures exaggerated Pks for fighter against fighter "ops" although 27% was better than the Pks in Vietnam.[8]

Air strategy on either side was non-existent. Both sides supported the army and used low-level operations to avoid radar coverage. Neither side had formalized air-ground coordination with their

armies, so close support was not as effective as it could have been.

SUMMARY

Other than the comparisons of weapons systems, there was little to learn fighter-wise from the Kashmir Conflict. Strategy and tactics appeared aimless. Neither side won.

Indo-Pakistan War 1971

INTRODUCTION

The cease-fire in 1965 was a timeout and both sides rebuilt their armed forces. When East Pakistan broke away from West Pakistan in 1971 and India supported the East (Bangladesh), war erupted again.

Below: The PAF had 100 F-86Fs when the Kashmir Conflict began. The F-86s did well against Hunters & Mysteres but had trouble with the Gnat.

AIR ORDER OF BATTLE

On the eve of war in 1971, the PAF had a mixed bag of aircraft:

• 40 F-86F Sabres

• 90 Mk-6 Sabres

• 70 F-6s (Chinese made Mig-19s)

• 20 Mirage IIIEs

• 7 F-104A/B Starfighters

• 16 B-57 Canberras

• 4 T-33As [9]

Again the IAF was bigger, as its own industry was supplemented by British and Soviet imports to 33 frontline squadrons of roughly 16 aircraft each. On the eve of the war, IAF strength was as follows:

o Mig-21PF and Mig-21FL -- seven squadrons

• Gnats -- seven squadrons

• Su-7 -- six squadrons

• Hawker Hunter -- six squadrons

• Mystere IV-A -- two squadrons

• Canberra Bombers -- three squadrons [10]

In addition, the small Indian Navy had 35 Sea Hawk fighters and the small aircraft carrier INS Vikrant. India had expanded its radar coverage extensively since the 1965 Kashmir conflict and had about 20 batteries of SA-2 SAMs emplaced around vital targets.[11]

The Indians had an objective to split Pakistan and supported the Mukti Bahini revolt against West Pakistan. On December 3, the PAF attempted an Israeli type preemptive air strike on Indian airfields. The IAF had hardened its airfields and these attacks were largely ineffective as there were no targets other than the runways and taxiways[12] (note that almost every air force in the

world went to hardened shelters after 1967 [except the USAF who only slowly adopted them in Europe] -- no effective air attacks were mounted against this type of defensive airfield hardening until 1991 in the Persian Gulf).

AIR WAR AND ANALYSIS

The India-Pakistan War was the second war of the supersonic era (Vietnam was the first) where it was again proven that high altitude Mach 2 capabilities were relatively ineffectual when operating at low altitude and when carrying external stores. The extremely high drag rise at Mach 1 limits the number of aircraft that can exceed the speed of sound at low altitudes and externally carried stores raised aircraft drag to the point where supersonic flight was not possible.[13]

What appeared again as truly important was the ability to maneuver without losing speed. This capability dictates a high power-to-weight ratio. It also appeared to be important that this ratio be determined with military power without the prolonged use of afterburner(due to range considerations and providing a massive IR target). Given these facts, it is not surprising that the Folland Gnat did well as did the Mk 6 Sabre with its big engine (7,300 # thrust Orenda) which could outmaneuver the Gnat. Another aircraft in this category was the F-6 (Mig-19) which also did well.[14] It appears to me that most of the pilots of this war were not trained in the proper use of speed and maneuvering in the

vertical which requires a great deal of experience and the highest level of aircombat training. It, therefore, should be a concern about which type aircraft a country should choose for its air forces.

India admitted 12,000 casualties including 3000 dead. Independent sources put her losses at 20,000. India claimed 94 aircraft kills, 54 Sabres, nine F-104As, six Mirage IIIs, and 25 other aircraft while sustaining 54 losses of their own.[16]

Pakistan claimed 106 kills: 32 Hawker Hunters, 32 Su-7Bs, 10 Canberras, nine Mig-21s, five Mystere IV-As, five HF-24 Maruts, three Gnats, and 10 others, for a loss of

only 25 Pakistani planes. Independent sources put Pakistani losses at 40 from all causes.[17]

Indian airfields were not only hardened, but heavily defended with SA-2s and AAA. Pakistan admitted five losses to AAA and SAMs around the airfields for the entire war. This appears low as we know that three B-57s were lost on one night alone.[18]

The Indian close support system was vastly improved over the 1965 conflict and coordination of air to ground was much improved. New weapons systems were also introduced. Counter-air airfield missions in

particular were helped by these new weapons. For example, Tezgaon airfield in East Pakistan was crippled by rocket boosted (similar to the Israeli "dibber" bombs) 1,100 pound bombs which created large craters in runway foundations that were difficult to repair. The Indians also developed a munition that created an entrance crater burrowed under the runway and then exploded creating a undermining tunnel and explosion crater.[19]

The new Indian fighter-bomber, the Soviet Su-7B proved very vulnerable. Out of 145 aircraft, 32 Su-7s were lost. The best Pakistani fighter-bomber, the Mirage, was very

effective. The Indians claimed six Mirages were shot down, but the Pakistanis said that none were. The high air-ground sortie rate of over 2,800 by the Pakistanis seems to support them.[20]

The Indians claimed nine F-104As had been shot down during the war. The Pakistanis owned up to losing only three. When it was pointed out that Pakistan only had seven F-104s to start the war, the Indians maintained that Jordan had reinforced Pakistan with 10 F-104s. Jordan admitted after the war that the Pakistanis had in fact been reinforced. The much awaited F-104/Mig-21 duel took place and the Mig 21 Fishbed won.[21]

This did not surprise me. When flying F-100s in Europe I had very little luck with F-104s, but when I checked out in the F-4, it was a new ballgame. In the F-4, I was as fast as the F-104 and could turn better. In the "Hun", all I could do was out turn the Starfighter. If the '104 kept his Mach up and used the vertical the Hun was easy meat, but the F-4 could take both away from the '104. When flying the Phantom, I used to look for F-104s. As the Mig-21 could out turn and out climb the F-4, the results in the Indo-Pakistani war of Mig-21 vs. F-104s appear viable to me.

The Pakistani F-6 and Mk. 6 Sabres proved to be admirable dogfighters. On the IAF side the Gnats again were superb. In spite of the large number of aircraft armed with missiles, most of the kills were still with the gun.[22] The most effective missiles were still the IR Sidewinder and Atoll. These early missiles could not handle the high G environment of the dogfights. Interestingly, the IAF customized the Soviet Mig-21PF to a Mig 21FL configuration which as first priority added two 23 MM cannon to the gunless 21PF.[23]

Left: *How the F-104 would perform against the Mig-21 was of great interest in the west. The Mig-21 generally prevailed in individual combats with the F-104.*

SUMMARY

The Kashmir and Bangladesh Indo-Pakistani Conflicts proved to be an operational test of then current East and West fighter inventories. All of the lessons we had learned in Vietnam were repeated:

• Supersonic aircraft had no advantage over lighter transsonic fighters when operating at low altitudes or when laden with external stores, particularly, if the transonic fighter had a high power-to-weight ratio.

• AAMs must be able to be successfully launched in high G fights to be effective. One must be able to fire missiles from any target aspect - not just the tail.

• Soviet weapons and aircraft interdicted airfields in East Pakistan by destroying runway foundations with specially rocket boosted bombs.

• Hardened shelters have changed the name of the game in attacking airfields. The shelters must be successfully attacked to destroy aircraft. The USAF has only hardened its airfields in Germany and Great Britain.

• The most successful fighters of this era were those with the highest thrust to weight ratios.

Below: *The Mig-21 proved formidable for the IAF. Shown is the 2-seater Mongol version.*

168

NOTES

1. Lon O. Nordeen, *Air Warfare In the Missile Age*, Smithsonian Institution Press, Washington, D.C., 1985, p. 79.
2. Op. Cit.; pp. 79-83.
3. Ibid.
4. Op. Cit.; pp. 90.
5. Ibid.
6. Ibid.
7. Op. Cit.; p. 91.
8. Ibid.
9. Op. Cit.; p. 93.
10. Op. Cit.; pp. 93-95.
11. Ibid.
12. Op. Cit.; p. 96.
13. Iain Parsons, *The Encyclopedia Of Air Warfare*, Salamnder Books, New York, 1975, p. 194.
14. Ibid.
15. Nordeen, Op. Cit.; p. 103.
16. Ibid.
17. Ibid.
18. Op. Cit.; p. 104.
19. Ibid.
20. Op. Cit.; p. 105.
21. Op. Cit.; p. 107.
22. Ibid.
23. Op. Cit.; p. 94.

Above: *The Argentine Dassault Entendard equipped with Exocet missiles proved to be a viable weapon against British Warships in the falklands War.*

CHAPTER 10

THE AIR WAR IN THE FALKLANDS (MALVINAS)

INTRODUCTION

Argentine Armed forces landed in the Falkland Islands at Port Stanley and Grytviken on South Georgia Island on April 2, 1982.[1] The British Government, under Prime Minister Margaret Thatcher, immediately began to prepare for war. The ownership of these islands had been argued for many years. When Argentina resolved to solve the problem by military force, Great Britain and her aggressive Prime Minister selected the military option for recapture of this small portion of Britain's remaining Empire. The "Jumpjet" (V/STOL) aircraft carriers Hermes and Invincible with their escorts departed England on April 5, 1982 for the South Atlantic. Onboard were 20 Sea Harriers vertical/short take-off and landing (V/STOL) fighters.[2]

The U.S. Secretary of State Alexander Haig began his version of "shuttle diplomacy" trying to stop the coming conflict. The British, already preparing to use the military option, declared a 200 mile radius "maritime exclusion zone" around the Falklands while Haig was in the air on April 7, 1982.[3] He met a quiet, firm position in London, and a more emotional reception in Buenos Aires. The Argentines would withdraw if the British would, but the British must leave Argentina's flag flying.

There was no immediately available Fleet Air Arm reinforcement for their Harriers as they had only purchased 28. There were, however, RAF squadrons of GR 3 Harriers built for ground attack that were not designed to operate from carriers. A crash program was put into operation to convert these aircraft to Sea Harrier configuration, particularly the ability to carry and fire the AIM-9L Sidewinder missile. The Ferranti navigational system had to be modified and the aircrews trained in "Jumpjet" operation.[4] Six of these GR-3s were shipped on the Atlantic Conveyor to the Falklands. Four more GR-3s were flown down to the Falklands from Ascension Island later in the War.[5]

The V/STOL Harrier was the key to air superiority for the British. This was the first combat for V/STOL aircraft. The world watched closely to see how these highly maneuverable, sub-sonic fighters would do against the supersonic Mirage IIIs of the Argentine Air Force. Remember it was the requirement to bring to bear a gun aligned with the longitudinal axis of the aircraft which brought about the need for maneuverability in the first place.[6] How would V/STOL/vectored thrust capabilities change this equation? Until the Falklands War, we had no real combat evidence.

The real question was the effect of a combination of vectored thrust with all aspect IR missiles such as the Harrier/Aim-9L combination. The Mirage/Daggers (the Dagger was an Israeli adaptation of the Mirage III — throughout the rest of this book the Dagger will be referenced as a Mirage) of the Argentine Air Forces were armed with earlier versions of the AIM-9 and French Matra radar missiles. That a VTOL aircraft had the advantage of maneuverability was not questioned, but how would it handle the mach 2 Mirage in an all-out fight? No one knew. Most thought the Mirage would prevail (Argentina had even turned down an offer to purchase Sea Harriers)[7]

In mitigation, the Mirages were forced to operate at a maximum combat range which limited their use of afterburner and their overall speed and energy advantages. A key was that these aircraft were not air refuelable. In addition, the modern SAM systems on the British Ships forced the Mirages down to low altitude where fuel consumption was exceedingly high.

Order of Battle
(May 1, 1981)

Argentine Air Force (Fuerza Aerea Argentina [FAA]):

- 11 – Mirage IIIEs
  60 – IA-58 Pucaras

- 34 – Daggers
  7 – C-130s

- 46 – Skyhawks 2 – KC-130s[8]

- 6 – Canberras[9]

Argentine Naval Air Force (Avacion Naval Argentina [ANA]):

- 10 – Skyhawks

- 5 – Super Etendards

- 10 – Macchi M-339s[10]

Below: *The VSTOL Harriers proved to be highly flexible in the falklands. They operated from "jump" carriers in weather in which ordinary carrier forces could not operate. Armed with Aim-9L all aspect sidewinders they were deadly to Argentine A-4s,Mirages & Daggers.*

Royal Navy:

- 20 – Sea Harriers (reinforced by 8 more Sea Harriers on 18 May)[11]

RAF:

- 6 – GR 3 Harriers used mainly as fighter-bombers[12]

On Ascension Island:

- 55 & 57 Squadrons – Victor tankers

- 44, 50, & 101 Squadrons – Vulcan bombers

- 120, 201, & 206 Squadron – Nimrod maritime reconnaissance aircraft (15 May)

- C-130 Transports (16 May)[13]

As the above lists indicate, 20 Sea Harriers gained air superiority from day one in the Falklands. After the first air battle on May 1, 1982, the Argentine Fighters never really contested the dominance of the Harrier and their AIM-9L mis-

siles. Interviews with the Argentines indicate they did not try to attack the Harriers because with the AIM-9L and the Harrier's maneuver capability, they felt it was futile.[14]

The Lineup

- FAA Dassault Mirage IIIE; 1 x SNECMA Atar 9C, 9,436 #s dry, 13,230 #s AB; 745 MPH at sea level, Mach 2.2 at 39,000 ft; 2 x 30 MM, 2 x AIM-9B Sidewinder, 2 x Matra/Nord missiles.[15]

- ANA Dassault Super Etendard; 1 x SNECMA Atar 8K-50, 11,025 #s, no AB; 745 MPH at sea level; 2 x 30 MM, Magic AAM missiles, bombs, rockets, 1 x Exocet AM.39 AGM.[16]

- RFAA British Aerospace Sea Harrier; Rolls-Royce Pegasus, 21,500 #s, no AB; 598 MPH at sea level, 607 MPH at 36,000 feet; 2 x 30 MM, 5 x AIM-9L Sidewinders, rockets, bombs (HE & Cluster).[17]

- RAF British Aerospace GR-3 Harrier; Rolls-Royce Pegasus, 21,500 #s, no AB; 598 MPH at sea level, 607 MPH at 36,000 feet; 2 x 30 MM, 5 x AIM-9L Sidewinders, rockets, bombs (HE & Cluster).[18]

Lineup Analysis – The FAA Dagger was an Israeli export version

of the IAF Nesher, a fighter-bomber version of the Mirage III (Mirage 5 in France). Engine, airframe, etc. were almost identical to the Mirage III. The fuselage was extended slightly with hard points on the wings and could carry 8,000 #s of ordnance.[19]

The Sea Harrier and Harrier were nearly identical (Sea Harrier had a shorter nose, higher seating for the pilot) except for those things needed to operate aboard ship. These included lash downs for outrigger landing gears, holes to drain salt water from the airframe, changes to nosewheel steering, and changes to the engine nozzles. Operational changes were the fitting of AIM-9Ls (previously requested by the RAF for months), installing compatible transponders and IFF, and modifying the inertial systems to be used at sea (aboard ship).[20]

The AIR WAR

In the May 1 fight, the Mirages and Sea Harriers passed head-on, and the FAA Mirages tried to scissors with the Harriers.[21] A scissors is not the best maneuver for fighting an aircraft with VTOL capabilities as vectored thrust will allow the Harrier to decelerate and almost rotate in place. The scissors maneuver is an effort to get your opponent to overshoot and allow you to get on his tail[22] – an impossible maneuver for a Mirage against a Harrier if the Harrier pilot can still breathe. In other words, this was the worst possible type of maneuver for the FAA pilot to attempt. The only possibility the Mirages had was to gain and maintain excessive energy, dive and

zoom, and achieve an advantage using their extra energy. This was tricky because they were operating at max range and could not use burner, but by converting altitude it could have been done. Coordinated supersonic attacks were the only way to go.[23]

On the basis of this one mission, where they used incredibly bad tactics, the FAA and ANA gave up attempts at air-to-air combat entirely, and adopted hit and run anti-ship tactics. It would appear that a massed and simultaneous attack against the Harriers and British shipping would have had a much better chance of success. Why the Argentines did not use mass and their overpowering numbers is unknown. One reason may have been the long-range efforts required with only two tankers. The Mirages and Daggers had no refueling capability so were limited and with only two C-130 tanker aircraft, the Skyhawks and Etendards were limited as well.[24]

It must be remembered however, that the invasion was an Argentine initiative and it took the British a month to reach the Falklands. Before they invaded, the Argentines should have developed a better air-refueling capability, and planned to build massive and suitable airfields in the Falklands during the time it took for the British to react and get to the South Atlantic. All of this should have been foreseen and planned prior to the conflict.

As it was, the Argentines still inflicted a heavy cost on the British. This was the first war with

all-out attacks by an air force against shipping since WW II. Air-to-ship Exocet missiles were effective in sinking two British ships (the Argentines only had five). The real surprise however, was that the low-level iron bomb attacks were so effective.[25] Free-fall bombs sank four British ships (two frigates, one destroyer, and one assault ship) and damaged 10 others. Less than half the bombs detonated.[26] (Ethell and Price estimated three quarters malfunctioned.)[27]

The British themselves estimate they would have lost six more ships had those weapons detonated.[28] All of the malfunctioning bombs were British 1000 pound bombs and there were evidently two problems with the malfunctions:

• Releasing the bombs too low for the fuses to arm. The bombs needed to be released above 200 feet to arm. Where British SAMs were active, the FAA and ANA had to fly lower than that. Why the Argentines did not know this or realize it during operations is unknown.[29]

• The fusing delays were too long. The bombs delivered at 500-600 knots often passed all the way through the thin skinned British ships before detonating. Others, that did not detonate at all, might have failed due to fuse distortion before the delayed detonation could occur. The delay is necessary to allow a fast moving low-level fighter to escape the blast of its own bomb.[30]

The Argentine fusing problems

smack of a combat "on-the-spot" solution without proper analysis by experienced weapons personnel in the FAA. They were forced down by British missiles and evidently did not know their weapon's fusing required higher altitudes than they were using. The time delay errors could not be known without extensive tests against actual shipping targets. In mitigation for the Argentines, low-level attacks and skip-bombing had not been used against shipping since WW II when there were no shipboard SAMs.

The use of "slick" weapons at low altitude was discarded in the USAF prior to the Vietnam War. We could not get our fuses to function properly. Most set for delay, detonated prematurely above 400 knots delivery speeds. This combination was deadly to the bomber from the blast of its own bombs. Longterm delays (10-12 seconds as used in the Falklands) seldom performed correctly. USAF ordnance personnel solved these problems by developing the retarded ("Snakeye"- where metal speed brakes open) family of weapons which provided arming time and allowed the aircraft to escape. This weapon system was not foolproof as the petal type speedbrakes would sometimes fail and the bomb would go slick. We often dropped in pairs watching each other's bombs. If a bomb went slick, we pulled straight up to get out of the slick bomb's frag envelope as it flew just below you after release. We also never dropped the weapon running perpendicular to the FEBA (never a good idea if it can be avoided) where a miss long (a slick bomb) could cause friendly casualties. Incidentally, Snakeyes have disappeared from the USAF inventory and been replaced with a parachute-retarded bomb.

Perhaps aware of their fuzing problem, the ANA used Snakeyes against HMS Ardent with devastating effect.[31] Had all the attacks been with Snakeyes or their equivalent, the outcome might have been very different. All of the Snakeyes detonated as advertised and Ardent sank.[32]

This was also a test of the first air-to-ship missiles. Only the Super Etendards were equipped with the Exocet anti-ship missile and these air-to-surface missiles were brand new. Without technical support from the French manufacturers, the ANA had a very difficult time maintaining these new Etendard/Exocet systems. The ANA launched only six Exocets, one of those from the

Below: *Pictured is an Aim-9 missile with and without head cover in place. The seeker head below the cover determines the Aim-9L's all aspect capability. It is effective from every angle even head-on, due to a cooled-head seeker.*

Above: *The Argentine Navy's A-4s could have been decisive if they all had been armed with low-level retarded ordinance (U.S. "Snakeyes"). As it was these air-refuelable attack aircraft sank several British warships.*

shore. Three missiles scored hits and sank *Sheffield,* and *Atlantic Conveyor* and damaged the *Glamorgan.* Argentine sources still claim one aircraft-launched Exocet hit a British aircraft carrier.[33]

ANALYSIS

The efficacy of naval forces in a missile environment bears some study. The Argentine attacks were determined and pressed home with great courage. They were also conducted with a few modern systems for their time (Exocet). Yet they were relatively successful against a relatively modern navy. That leads to some speculation. If the Argentines had possessed modern ECM equipment, had used sound principles and their numerical superiori-

ty against the score of Harriers available, what would have been the result? Or what would a determined attack with modern missiles and aircraft do against a U.S. Navy carrier task force today? Does the USAF and USN have sufficient stocks of low-level anti-ship ordnance suitable and fused for skip-bombing type deliveries? If our missiles fail to score against an enemy fleet, are our aircrews trained to go in low? Those questions need answers.

AIR REFUELING

Air refueling played a crucial role in the operations of both air forces. With British operations, air refueling was required from England to Ascension Island for Vulcan heavy bombers, Nimrod maritime reconnaissance aircraft, and C-130 transports. It was required for these same aircraft to cover the 3,000 plus miles from Ascension to the Falklands. The Nimrods were used for reconnaissance and the Vulcans flew two long-range bombing missions against Port Stanley airfield from Ascension. Early on, the Victor tanker aircraft were used for recon, flying over 7,000 miles by swapping the fuel in four tankers so that one could complete the sortie around South Georgia Island.[34]

Argentine operations required refueling, where possible, as they were operating at maximum range and were forced to offload ordnance and carry maximum external fuel (e.g., Rio Grande air base was 437 miles from Port Stanley and Comodoro Rivadovia 596 miles).[35] A-4s and Etendards often refueled both outbound and inbound to the Falklands and time had to be minimized in the target area.[36] As pointed out earlier, the Mirages and Daggers had no air-refueling capability and there were only two C-130 tankers, so scheduling tanker support was difficult at best.[37]

Argentina capitulated on June 14, 1982 and the fighting ended.

SOME ARGENTINE FAILURES

It is hard to believe that Argentinian Leaders thought the British would react as strongly as she did. However, if one chooses aggression one better be prepared for the consequences. The FAA and ANA broke down in several areas:

• There was no overall air battle plan to achieve air superiority and defeat the British.

• FAA/ANA reconnaissance and intelligence was poor throughout the conflict. Fighter pilots were on their own for ship locations except for what guidance the TPS radars could give them based on where they last saw returning Harriers. Also, on occasion, they received troop reports about ships offshore.

• There was apparently no analysis of British weapons and their strengths and weaknesses.

• It is incomprehensible that in 1982 any air force should possess no plan for ECM support or IR defensive devices to counter the AIM-9L, Sea Dart, Rapier, and other British missiles.

SUMMARY

By any standard, the Falklands War was a small war, but its implications were huge. The first was how did a V/STOL aircraft like the Harrier do as an air superiority fighter? As I have tried to indicate, the Falklands War may not be a good indicator. Neither the FAA or the ANA made a concerted effort to fight an air superiority battle. The one FAA attempt was made with inane, if not insane, air-to-air tactics. The American AIM-9L Sidewinder AAM dominated the air battle. For 26 launches, there were 19 confirmed kills (a 73% Pk). Ethell makes the point that three of these were double attacks where the "Brit" pilot did not get an immediate kill and launched a second missile. On two occasions, the target then disintegrated before the second missile struck or exploded on the second missile's arrival. One double launch did not result in a kill. Ethell used missile engagements versus kills; therefore, according to him the proper Pk should be 19 kills for 23 engagements or 83% (that's a new definition of Pk). Either way the Sidewinder was dominant. Unlike any other war I know of, the Harrier pilots turned out to be conservative in their kill claims. When Ethell interviewed Argentine pilots, he discovered every claim of a kill was valid and two "possibles" were actual kills.[38]

Incidentally, in 1995 the Soviet AAM-11 missile is conceded to be the best IR air-to-air missile operating today. The results in both the Falklands and the Bekaa Valley indicate the best missile gets air superiority.

As indicated, the attack of shipping by modern/semi-modern jet aircraft needs to be analyzed by U.S. strategists. We have not had a determined air attack on our shipping since WW II. On the offensive side, I am not sure we have an inventory of dumb retarded weapons for low-level shipboard attacks such as those conducted by the Argentines. A careful analysis should be done considering missiles, smart bombs, and low level attack requirements with an emphasis on fusing and armor penetration. In my opinion, the Argentine Air Forces would have done extensive damage to U.S. ships just as they did the British, if we fought defensively as the Brits did. Hopefully, we would have taken out the Argentine bases on the mainland as the British should have.

The British Navy in 1982 had no adequate defense for their shipping against the Exocet missile. Its fire and forget technology made it exceedingly difficult to stop. Our shipboard defenses need to be reviewed versus this threat considering its upgrades since.

This study should also consider ECM and stealth requirements to penetrate ship defenses. A stealthy Exocet should be interesting.

Again no one targeted air-refueling resources which are critical to modern air war.

The RAF had been unable to sell their legislature on an AIM-9L installation until war was declared. Its obviously endemic in democracies during peacetime to be unprepared (WW I, WW II, and Korea come to mind).

Like the U.S. in Korea and Vietnam, the British chose to risk their men rather than widen the war to the Argentine mainland. The FAA and ANA bases were very vulnerable, but Britain limited their attack.

The war was interesting, hopefully its lessons will not be forgotten like so many others.

NOTES

1. Bryan Perret, *Weapons of the Falkland Conflict*, Blanford Press, Poole, Dorset, 1982; p. 9.
2. Jeffrey Ethell and Alfred Price, *Air War South Atlantic*, MacMillan Publishing Company, New York, 1983, p. 2.
3. Max Hastings & Simon Jenkins, *The Battle For The Falklands*, W.W. Norton Company, New York, 1983, p. 106.
4. Ethell, Op. Cit.; p. 5.
5. Op. Cit.; p. 194.
6. Bill Gunston & Lindsay Peacock, *Fighter Missions*, Crown Books, New York, 1989, p. 46.
7. Perret, Op. Cit.; p. 76.
8. Lon O. Nordeen, *Air Warfare In The Missile Age* Smithsonian Institution Press, Washington, DC, 1985, p. 193.
9. Ethell, Op. Cit.; p. 7.
10. Nordeen, Ibid.
11. Ethell, Op. Cit.; pp. 4-5.
12. Ibid.
13. Op. Cit.; p. 194.
14. Ethell, Op. Cit.; p. 175.
15. Perrett, Op. Cit.; pp. 76-77.
16. Op. Cit.; pp. 77-78.
17. Op. Cit.; p. 75.
18. Op. Cit.; p. 75.
19. Green & Swanborough, *The Complete Book Of Fighters*, Smithmark Publishers, New York, 1994, pp. 155-156, 302.
20. Ethell, Op. Cit.; pp. 4-5.
21. Op. Cit.; p. 196.
22. Robert L. Shaw, *Fighter Combat Tactics And Maneuvering*, Naval Institute Press, Annapolis, MD., 1985, p. 82.
23. Ethell, Op. Cit.; p. 37.
24. Nordeen, Op. Cit.; p. 193.
25. Nordeen, Op. Cit.; p. 201.
26. Ibid.
27. Ethell, Op. Cit.; p. 183.
28. Nordeen, Ibid.
29. Nordeen, Op. Cit.; p. 183.
30. Ethell, Op. Cit.; p. 65.
31. Ethell, Op. Cit.; p. 92.
32. Nordeen, Op. Cit.; p. 198.
33. Op. Cit.; p. 201.
34. Ethell, Op. Cit.; PP. 14-16.
35. Op. Cit.; p. 194.
36. Ethell, Op. Cit.; p. 10.
37. Nordeen, Ibid.
38. Ethell, Op. Cit.; pp. 175-176.

Above: *The Su-25 "Frogfoot" was the Soviet A-10 in Afghanistan. It was a formidable air to ground aircraft, but proved vulnerable to all-aspect ground-to-air "Stinger" missiles.*

CHAPTER 11

SOVIET EXPERIENCE IN AFGHANISTAN

INTRODUCTION

The invasion of Afghanistan by the USSR is of interest to a study of fighter warfare for several reasons. It closely parallels U.S. experience in Vietnam and provided some insight into how to pacify a dedicated, nationalist people using modern weapons systems and troops.

Tactical fighters and masses of helicopters were deployed to Afghanistan by the Soviets and, like the U.S. in Vietnam, they were unsuccessful. Our interest lies in the answers to these questions: Why were they unsuccessful? How did they employ fighters? What lessons did they learn? Can we learn anything from their experience?

BACKGROUND

Afghanistan has been fought over as a route from Iran (ancient Persia) to India for thousands of years. Alexander's Phalanxes marched through this mountainous country to meet destiny on the Indus. In more modern times, Great Britain and Imperial Russia/the USSR have struggled over control of this centrally located Asian nation. Britain went to war in Afghanistan twice, both times in the 19th Century, and during the 20th Century, she dueled with the Russians for influence there until 1947. From that time on, it was considered a Soviet sphere of interest and Soviet aid of

all types was offered and accepted. Two clandestine Afghan Communist parties were established as Afghanistan became a Constitutional Monarchy in the 1950s. These parties the Khalq ("Masses") and the Parcham ("Banner") were united on paper in 1965 into the Peoples Democratic Party of Afghanistan (PDPA). [1]

Prime Minister Mohammed Daoud led a coup that took over the country on July 17, 1973. He struck while the King was out of the country. The coup was bloodless. He rewarded his PDPA communist allies with several cabinet positions and his policy counted on close alliance with the Soviet Union. [2]

Within a year however, Daoud had become disenchanted with the Communists and removed five from his cabinet. He then began to steer a more middle of the road policy, accepting a major loan from the Shah of Iran, his neighbor to the west. Much of his change was due to Islamic unrest about Godless communism in the government.

As tensions heightened, a Communist technician was murdered and panic ensued. Daoud rounded up the Communist Party (PDPA) members, but failed to clean them out of the military. On April 28, 1978, Daoud was overthrown by a military coup and he, and all of his family, were summarily executed. The coup, in

contrast to the bloodless one of 1973, cost about 2,000 other lives as well. [3]

THE KABUL REGIME

The new communist regime turned out to be repressive and met resistance from the traditional Islamic Afghans. The resistance called themselves the Mujahideen (fighter for the faith). The Communists feuded among themselves and the Soviets supported one faction against the other. The resistant Mujahideen turned out to be a formidable opponent. Afghans have a tradition of successful guerilla warfare and these mountain people are fiercely religious and independent. The Mujahideen were brave, dedicated fighters. [4] (Parallels with Vietnam are inescapable.)

On December 27, 1979, the Russians invaded in support of the Kabul Regime. The U.S. reaction was "swift and decisive." President Carter boycotted the Olympics in Moscow and put in a grain embargo which forced the Russians to buy Canadian wheat. [5] One wonders what he would have done if they had invaded Pakistan too. Cancel the World Series?

THE SOVIET INVASION

The Soviets committed their forces on a low intensity scale in that they wanted to hold their casualties to a minimum while maximizing those of the Mujahideen. The Soviets

181

invaded to put some backbone in the Kabul Regime pro-Communist troops who had been reduced by defection to only 20,000 troops. They found great difficulties in attempting to control Kabul and even more trouble with the Mujahideen in the field. The Soviets met strikes and civil unrest with gunfire and massacred thousands. All-out resistance appeared all over Afghanistan. The Soviets launched their version of search and destroy operations only to take heavy casualties from the fierce Mujahideen. Several changes in tactics then took place.

First the Soviets went to light infantry operations (1981) and began training Kabul Communists to take over the fighting. [6]

Then in 1982, after the Kabul Regime troops proved unreliable, the Soviets reverted to large unit operations which were strongly resisted by the Mujadiheen. By the end of 1982, the Soviets had adopted an Air War Strategy very much like the U.S. forces in Vietnam.[7]

THE AIR WAR

The Soviets began to extensively bomb cities and suspected Mujahideen positions. Unlike U.S. operations, Soviet forces conducted de-population missions against resistance population centers. Fighter bomber attacks throughout the war were designed to minimize Soviet losses and casualties.[8]

The Kabul Regime began to depend on the heavy Soviet firepower (artillery and airpower) in all of their operations (shades of the ARVN in Vietnam). The Soviets quickly found that in a third world country with rough, hazardous ground everywhere and few roads, the helicopter was the only way to achieve mobility.[9] The use of "Choppers" required air superiority. In 1986, the Mujahideen found a way to challenge Soviet air superiority when the U.S. CIA and other Allies supplied them with Stinger and Blowpipe missiles.[10]

The introduction of the Stinger was a real turning point of the war as the cooled seeker heads and long range of the missiles were deadly to both the feared, heavily armed, and armored Hind helicopters and the Soviet fighter-bombers, particularly the Soviet Frogfoot (the Russian equivalent of the U.S. A-10 Warthog).[11] Until the arrival of the Stinger and Blowpipe, the Soviet SA-7 Grail and RPG-7 rocket launcher had been the best defense against these heavy-hitting systems. The sensitive seeker-heads of the Stingers were immune to many of the counter measures for the SA-7 and there were no counter measures for the wire-guided Blowpipe. The speed and long range of the all-aspect Stinger kept the casualty sensitive Soviet fighter-bombers at very high altitude (often over 20,000 feet).[12] Helicopter operations were still effective after the appearance of the Stinger, but not nearly so decisive. The Mujahideen had an answer for their previously overwhelming firepower and it could restrict Soviet helicopter access for the first time.[13]

When the Kabul Regime troops lost the support of the Soviet helicopters and fighter bombers, they were no match for the Mujahideen in ground combat. In May 1988, with the Soviet Union disintegrating

at home, President Gorbachev decided to withdraw his forces from Afghanistan.

SUMMARY

By 1988, the Soviets had ten aircraft squadrons in Afghanistan and 15 more engaged from the USSR. From 30-50 regiments of helicopters were employed with 275 helicopters and over 300 just across the border in the USSR.[14] Included in the inventory was some of the VVS's best equipment: Mig-23 and Mig-27 Floggers, Mig-21 Fishbeds, Su-17 Fitter-Ds and Su-25 Frogfoots (Frogfeet?). The mix of helicopters included Hinds, Hips, Hooks, and Halos–most were Hips and 25% Hinds.

In addition to the VVS forces, the Kabul Regime had about four Regiments of fighters including: Mig-21 Fishbeds, Su-7 Fitter-Ds, Su-7s. (Mig-17 Frescos and Mig-19 Farmers had been withdrawn by 1988.) Also included were 60-80 helicopters and 30-60 turbo-prop transports.[16]

Below: *Soviet "Hind" helicopters were armored against small arms, but shoulder fired missiles often forced them to skirt or stop operations.*

Soviet losses in Afghanistan were 13,310 killed and 35,478 wounded. The Kabul Regime is estimated to have lost over 200,000–many of which were deserters.[17]

The Soviets had lost approximately 1,000 aircraft by 1987. More than 80% of these were helicopters. About half of these helicopters were lost during operations while many of the others were destroyed on the ground by Mujahideen attacks.[18]

ANALYSIS

What were the lessons of Afghanistan? The same as Vietnam. Even with overwhelming strength, the two greatest military powers in the world were unable to subdue a dedicated, committed, third world country so long as there was a sanctuary close by and a source of external supply. Soviet air power had no inane rules of engagement such as we imposed on our flyers. Fighter-bombers were routinely employed against Afghan populations where the U.S. targeted only military facilities, but other than that, the employment was the same. It was proven again that it is impossible to interdict a guerilla force from its supplies with air power alone. The Stinger made life much more hazardous for Soviet helicopter pilots than our "Chopper" pilots in "Nam" and does not bode well for helicopter employment in the future without suitable countermeasures.

CONCLUSION

Two of the greatest nations on earth became involved in third world guerilla wars in the last 30 years. Both lost those wars. Air power was used extensively in both, but as we have seen in this course, the ability of air power to decisively affect guerilla war is limited, particularly, if the third world power is tough and dedicated.

Modern shoulder-mounted IR missiles can be deadly to helicopters and fighter bombers without effective counter-measures. Guerilla operations still give a lesser opponent the opportunity to prevail if air superiority can be limited in this way.

As the Mujahideen leader Armad Shah Massoud stated in Afghanistan, "There are only two things Afghans must have: the Koran and Stingers".

NOTES

1. David C. Isby, *War In A Distant Country Afghanistan: Invasion And Resistance*, Arms And Armour, London, 1989, pp. 15-25.
2. Ibid.
3. Ibid.
4. Ibid.
5. Gerard Chaliand, *Report From Afghanistan*, Penguin Books, New York, 1982, p. 84.
6. Isby, Op. Cit.; pp. 28-48.
7. Ibid.
8. Op. Cit.; p. 62.
9. Op. Cit.; p. 63.
10. Op. Cit.; p. 7.
11. Op. Cit.; p. 114.
12. Ibid.
13. Op. Cit.; p. 59.
14. Isby, Op. Cit.; p. 67.
15. Ibid.
16. Op. Cit.; p. 89.
17. Op. Cit.; p. 62.
18. Op. Cit.; p. 65.

Above: *The Lockheed F-117 Stealth Fighter was used to attack hard targets in Iraq at night. No one knew they were there until their guided smart weapons began to detonate. The F-117 ushered in a new era of warfare. No F-117 was hit during the war.*

CHAPTER 12

THE PERSIAN GULF WAR

INTRODUCTION

The Persian Gulf War was the first U.S. war since WW II that was fought to achieve a clear set of national objectives. We went to war to free The Sheikdom of Kuwait and to protect Saudi Arabia. We struck with the full force of the United States and her supporting Allies. Saddam Hussein and Iraq inexplicably allowed us the time to build up our forces until we were ready to strike. This time there was no gradualism. The political objectives were clear and understood, and they were attainable with the military forces committed to the task.

IRAQI INVASION

SADDAM HUSSEIN AND IRAQI ACTIONS

The Iraqis, and Saddam Hussein in particular, have a long history of resentment toward Kuwait. This resentment's modern history dates back to 1922 when present-day Kuwait and the borders of Iraq and Saudi Arabia were established by decree by a British Colonial Governor. Iraq resented the small 26 mile coastline on the Persian Gulf provided by that decree. (Saudi Arabia and Kuwait combined were provided hundreds of miles.)[1]

In recent years, following the Iraq-Iran War, this resentment intensified significantly as Iraq was deeply in debt as a result of the war, and dependent on the sale of oil to pay off these war debts. Kuwait was openly producing oil at reduced prices for world markets. Iraq was very upset by what they considered over-production by OPEC and Kuwait as Iraq needed prices in the $18/barrel range and the world price was at about $13 in June 1990 (the price was $20.50/barrel in January and Iraq held Kuwait largely responsible for the price reduction).[2]

The Iraqis demanded that the Kuwaitis stop over-producing oil and also demanded that they stop stealing Iraqi oil by parallel drilling. For years, Iraqi and Kuwait have argued about the ownership of the Romalia oil fields and several island and off-shore fields. Iraq demanded the return of this rich field and some of the islands. Saddam's government also asked for a direct subsidy of 12 billion dollars to offset Iraq's losses due to Kuwait's overproduction and demanded that Kuwait forgive Iraq's 10 billion dollar war debt borrowed during the Iraq/Iran War. The Kuwaiti Royal family resisted these demands.[3]

The Kuwaitis, used to the double-dealing often associated with world financial markets, ignored Saddam's threats and counted on the U.S. and world opinion to

---

Below: *French and British attack units used the Sepecat Jaguar for ground attack during Desert Storm.*

restrain Iraq. When Iraq insisted that Kuwait restrict her oil production, the Kuwaitis countered with reminders of Iraq's outstanding war debt.

According to one U.S. assistant secretary of state this drove Saddam Hussein wild.[4] When American policy supported Kuwait, Saddam may have misread America's determination to support policy with force. Many world leaders questioned our resolution and determination to act on stated policy after

war. Specifically, they did not believe that the United States had the willingness to go to war particularly where the loss of American lives was certain. They were very surprised by our actions after Pearl Harbor and shocked by our outrage, anger, and determination. For example, the tough Marines and "GIs" who landed on the Japanese-held islands in the Pacific during WW II were a real shock and surprise to the Japanese in the 1940s. These tough well-trained men, with their "take-no-prisoners"

make this mistake and often misread our capabilities and particularly our resolution and dedication. We are not an easy government for foreigners to read and understand. They mistake honest dissension and protest, for irresolution, and mistake civility for weakness. Unfortunately, 1960 Vietnam protests, our press, and our movies enhanced this irresolute image. We are a self-deprecating people and often criticize ourselves, our government, and its policies excessively. In this public self-flagellation, we often give people overseas the wrong ideas about our beliefs and motivations. Certainly the rich, indolent, and depraved society so often depicted in Hollywood films, and TV, projects the wrong image about our country to the world.

Above: *The Sukhoi Su-17 (20, 22) were the Iraqi Air Forces primary ground attack aircraft.*

our Vietnam experience. It appears that Saddam Hussein may have made this error.

U.S. policy makers really should not have been surprised by Saddam's mistake. In the not too distant past, several dictators made this same mistake. Hitler, Mussolini, and Tojo during WW II are good examples. They did not think democratic nations could produce the mental, physical, or spiritual toughness necessary to wage successful combat and all-out

attitude, did not fit the Axis powers' stereotype of the "soft", indolent, decadent, peace-loving Americans.

Ho Chi Minh and Giap went further in Vietnam. Following the dictates of Mao Tse Tung, they rightly calculated that western nations would tire and go home after years of "protracted conflict". Watching Vietnam and how the U.S. responded, nations around the world could have concluded that Americans were indecisive and unable to promote a national will. Obviously Saddam Hussein was one of these.

Today, other nations continue to

However to make and act on conclusions about America, and Americans is a complex task. Without deep study and investigation of our society and its depth and strength, it is at a foreign nation's peril to question our will and determination on any question. Any potential enemy should remember the words of President John Kennedy, "We do not want to fight, but we have fought before."[5] And prior to 1950, we, by any standard, had always won by imposing our will on our enemies.

Most of the world did not realize the level of frustration caused by the Vietnam War among the millions of Americans who fought there. Bush, Cheney, Powell, and Schwarzkopf were all determined that we had fought our last war of "gradual response". President Reagan had rebuilt our armed forces and we were determined to

Above: *Most of the "Smart" weapons delivered during Desert Storm were by F-111s.*

fight the next war to win. Our leaders saw to it.

In mitigation, Iraq's grasp of the real situation was probably over-complicated in the Persian Gulf by the U.S. government sending mixed signals on Kuwait. Even as U.S. officials indicated support to concerned Kuwatis, we were officially telling Saddam that we were not overly concerned about middle-eastern border disputes.

Few, if any responsible officials, foreign or domestic, expected the Iraqi invasion although every intelligence bureau knew Saddam had the capability and nearly everyone watched his buildup on the Kuwaiti border with concern. No one believed he would do it, or at least that he would invade and annex all of Kuwait[7] – least of all the Kuwaitis, who behaved abominably at an 11th hour attempt to avoid the crisis, although Iraq, Egypt, Saudi Arabia, and Jordan advised that Saddam was dead serious. Kuwait evidently thought he was bluffing. Tariq Aziz, Iraqi Foreign Minister, stated in an interview after the war that it was after this mideast summit that Saddam committed to invade.[8]

The invasion at 0200 Kuwaiti time, August 2, 1990 was swift and effective. A country about as large as the State of New Jersey was overrun within 12 hours. The invasion was led by six divisions of the elite Republican Guard. Two of these divisions with their armored brigades advanced to the capitol. Special Forces troops were landed by ship and helicopter. Armored personnel carriers landed by amphibious craft attacked the palace. Casualties were less than 100 persons on both sides. It was a professional performance by the Iraqi army–swift and complete.[9] By the time the ground attack phase of Desert Storm began, these forces and those in Southern Iraq had grown to 39 divisions: Thirty Infantry and Mechanized Infantry divisions and nine Armored Divisions.[10] According to the Pentagon this amounted to an estimated 590,000 men (reduced to 350,000 after the war) 5,750 tanks, 700 aircraft, and 15 ships.[11]

DESERT SHIELD

THE U.S., ITS ALLIES AND THE UN

As shocked and surprised as were

the U.S. leaders, the initial reaction in the U.S. was relatively mild. Members of the National Security Council were heard to joke that we had not lost the "gas station" just changed the name out front. U.S. leaders immediately announced there were no plans for a U.S. military reaction. But in just a few hours U.S. Policy hardened quickly. Assets were frozen and all trade suspended. The next day, Secretary of State Baker and Russian Foreign Minister Shevardnadze issued a joint call for international sanctions by the United Nations.[12]

By mid-afternoon August 3, President Bush announced at a joint press conference with Margaret Thatcher in Aspen, Colorado that the U.S. had not ruled out any options. The American press pointed out that the President had re-opened the bidding for the war option. Many thought this stand was brought about by the redoubtable Ms. Thatcher.[13]

Immediately following the invasion, President Bush dedicated his efforts into organizing an international coalition against Iraq. The dissolution of the USSR had given us the ability to work in the UN Security Council without fear of the

veto. Bush announced flatly that the aggression in Kuwait "would not stand." International sanctions were instituted on August 6 by UN Security Council resolution against Iraq and an embargo/blockade immediately began to be implemented.[14]

On August 8, the President announced to the nation that the 82nd Airborne Division and select Air Force units were being dispatched to Saudi Arabia to protect that country from possible aggression by Iraq. The buildup had begun. Their mission was to defend themselves, Saudi Arabia, and other American friends in the Gulf.[15] Before this mission was over, total Allied forces would equal 700,000 personnel, 1,746 aircraft, 3,673 tanks, and 149 warships.[16]

Unlike every President since WWII, President Bush set political objec-

tives for his military in the Persian Gulf, provided every means of military and diplomatic support, and then let them do their job. No President since Franklin Delano Roosevelt had that kind of moral courage. All of the mismanagement recorded in Korea and Vietnam, the impact of Nuclear Weapons on conventional warfare and American command and control, was sorted out successfully by the Bush administration. He, and his defense department finally did away with the "strategic defensive" and "Vietnam Syndrome"[17] that we had been saddled with for years.

In the end, he got very little credit for this incredible feat in our topsy

turvy world. President Bush's initial high approval ratings plummeted when the economy receded. More concerned about the domestic economy, the American people voted him out of office in 1992. The man who was voted in avoided military service during Vietnam. Such is the impact of a largely mismanaged recession after an incredibly successful war.

As admirable as the President's performance was in Desert Shield/Storm, and in the handling of the military, insufficient analysis has been done by most Americans on Saddam Hussein's objectives and his efforts to kill two birds with one stone with his invasion of Kuwait. It appears that Saddam wanted to cure his vast financial problems and to be recognized as the leader of the Arab world in one stroke. The record indicates that early into the crisis he never believed he would get into a shooting war with the U.S. and her Allies, but could become an Arab hero by defying the "Great Satan".

As outlandish as that may sound to the American mind, it was a shrewd assessment of pan-Arab feelings. The have-not PLO, and populations of Iraq, Jordan, Yemen, Iran, Pakistan, Syria, Egypt, etc. all were for Saddam. Fundamental Islamics blamed the U.S. for the wealthy, decadent behavior of the oil-rich Kuaitis (and the Saudis too). There was sympathy for Saddam throughout Islam.[18] In addition, Arabs in general blamed the U.S. for the presence of Israel and in particular for the insufferable Israeli settlements on the West Bank. Saddam, always empathetic and helpful to the PLO, intended to become an Islamic hero for the "Arab Nation". It appears it was his plan from the start if there was a "Second Gulf War" to force Israel into the war to destroy any U.S. coalition and unite the Arab world against the U.S. and Israel.[19]

The Arab coalition put together by Baker and Bush was key to Arab unity behind the Allied coalition. However, many Arabs and the fundamental Islamics were very uncomfortable with this uneasy alliance.

Left: *The Grumman F-14 with its long-range Phoenix missile was the primary U.S. Navy air defense fighter.*

THE U.S. AND ALLIED BUILD-UP

No one but possibly God, the Devil, and Saddam Hussein know why Saddam allowed us to build up our allied forces in Saudi Arabia after his invasion of Kuwait. If he had invaded Saudi Arabia, and interdicted the airfields, ports, and approaches to both the airfields and ports with his air force and followed up with his battle-hardened army, the campaign would have been quite different. There is little doubt he could have made our build up very difficult. If one rolls out a map of the mideast and eliminates the bases in Saudi Arabia, the strategic problems for the Allies are obvious. But he didn't.

Even if he had attacked after our first troops were airlifted in, the results looked untoward for the U.S. The 82nd and 101st Airborne divisions that preceded the major buildup referred to themselves as the "Speed bump".[20] If the Republican Guards had invaded with an all out armored attack, it is doubtful that our two light divisions and the Saudis could have held out, particularly if our bases in Saudi Arabia had been hit with all-out air simultaneously. It certainly would have been close.

Allied leaders were acutely aware of this danger and built up as rapidly as possible. By mid-September Allied forces were up to 100,000 and rapidly building. General Schwarzkopf, as Allied Commander told President Bush it would take 250,000 troops and overwhelming airpower to safely protect Saudi Arabia from invasion. He also noted it would take four months of buildup before this could be accomplished. The American plan was 250,000 troops by 1 December 1990. Schwarzkopf carefully stated to the President that this was a force with only a defensive capability.[21]

ALLIED AIR BUILD-UP

Unlike Korea and Vietnam, the USAF and USN had four fighter weapons systems equal to or superior to any others in the world. The F-14, F-15, F-16, and FA-18 could hold their own with any in the world. Iraq had only two aircraft in this league, the Mig-29 and Mirage F-1. Unlike Vietnam however, the allies had good IFF, superb on-board missiles, and good supporting radar coverage. Even Mig-29s and F-1s had no chance.[22]

Since Vietnam, USAF & USN aircraft had progressed to a quantum level far beyond anything anyone had ever seen before in engine power, maneuverability, aircraft control, and aircraft weapons systems. In addition, aircrew training in air-to-air combat had progressed to the graduate level compared to

earlier training. We had the best trained aircrews in the world due to "Red Flag", "Top Gun", and other Allied realistic training. The aggregate was the best-trained, best equipped fighter force in our history to take on the Iraqis.

ALLIED OPTIONS AFTER SAUDI ARABIA WAS SAFE

Early in October, Washington began to consider offensive options. Schwarzkopf was fit to be tied as all of the forces he needed to defend Saudi Arabia were not in place. He was very nervous that Washington might escalate to the offensive without adequate forces in place.[23]

Major General Robert B. Johnson,

Schwarzkopf's Chief-of-Staff, briefed the Pentagon on offensive possibilities on October 10. He called for a four-phased attack, the first three phases being air attacks, which were readily accepted, but the ground attack was heavily questioned. Johnson had begun the briefing by caveating his inability to adequately plan this new option as there had been insufficient time. The next day when he was questioned by the President and his staff, he indicated he had insufficient force to use a wide turning movement rather than a frontal attack. Johnson indicated 1 January to 15 February before they could be ready. Both the Pentagon and Scwarzkopf's headquarters began studying a more indirect approach (a wide "left hook"), particularly, force requirements, and the logistic requirements for such a move.[24]

DESERT STORM

THE AIR WAR

The Air War in the Persian Gulf proved to be something of a surprise to most people, many of them "experts". Prior to the beginning of the conflict, the predictions of massive casualties among Iraqi noncombatants from Allied air strikes dominated the media. Expert after expert predicted that Allied air would cause thousands of civilian casualties. Nor was the effectiveness of the allied airpower predicted. The experts thought that airpower would assist the ground war, but no one foresaw the accuracy, incredible efficiency, and decisiveness of the actual attack.[25] Perhaps the most surprised of all was the

Iraqi high command and Saddam Hussein.

Saddam Hussein really underestimated the effects of air power. The air assault of Desert Storm had to come as a rude shock. As the crisis began in 1990 Saddam was quoted, "The United States relies on the Air Force and the Air Force has never been the decisive factor in the history of wars."[26] Saddam had committed the "Mother of all blunders". The Persian Gulf War was decisively influenced by Allied air power to an extent never before seen by anyone, least of all the myopic Hussein.

The plan for the air battle in the Persian Gulf evolved from roots that began during the frustrations of the Vietnam War. Air officers like their ground contemporaries were really frustrated in Vietnam. Vietnam air power was misused from day one, and most of the junior officers who served there agonized over the misuse of air weapons, and the unnecessary losses those mistakes cost. They vividly remembered these mistakes: Lack of a clear, attainable national objective, rotten target selection by DOD amateurs, stupid rules of engagement, failure to destroy NVAF fighters and SAMs during their initial build-up, the lack of a gun in our air to air fighters, fighters with no maneuverability, and the use of tactical fighters as strategic bombers and vice versus.

Air Force officers were resolved to correct those errors and improve their equipment to attain strategic air superiority in any future war. The Reagan years allowed them to build and hone a magnificent air

Above: *Mig-23 was one of Iraq's mainstays for ground attack.*

arm: The F-4Gs, F-16s, F-15s, F-15Es, A-10s, B-1Bs, B-52s, EF-111s, E3B AWACS, hundreds of air-to-air missiles, ALCM cruise missiles, "smart" bombs, and extensive ECM/electronic jamming packages. All were ready. The Navy had complementary packages that included; A-7Es, AV-8As, F14s, A/F-18s, E2As, EA-6s and Tomahawk cruise missiles.[27]

The plan began months before as "Instant Thunder" (a takeoff on the gradual strategy in Vietnam termed "Rolling Thunder") within the bowels of the Pentagon air planning staff.

It identified five Iraqi "centers of gravity" as targets and was originally conceived by Colonel Jonathan A. Warden III. The plan was briefed to all of the senior commanders and modified extensively by Generals Horner and Glosson in Saudi Arabia. Unfortunately, Horner and Warden had a personality conflict during the briefing to Horner. Warden subsequently became persona non grata with the Air Force staff in Riyadh.[28] The final product was one of the most sophisticated air strike plans in history and beyond Saddam Hussein's wildest nightmares. The surprise, violence, weight, and precision of this attack was to set new standards for air war.

The plan targeted and attacked Warden's "Five Centers of Gravity": **Military/Civil Leadership** - Command, Control, & Communications (C3), and Internal Control Organs; **Key Production** - Electrical Power, Refined Oil Production, Nuclear, Biological, Chemical Weapons, and R&D Production; **Infrastructure** - Transportation Nodes, Railroads, and Bridges; **Population** - **Regime Support**; **Fielded Military Forces** - Air Defenses, Air Forces, Army Forces and Navy Forces.[29] This approach was different in that all five centers of gravity were hit simultaneously with the copious air power available to the coalition.[30]

It was Warden's belief that this plan (Instant Thunder) would win the war all by itself. It did not, but it came closer than any air campaign in history.

Enemy Air Forces

The seizing of air superiority appeared to be a tall order. Unlike the Koreans and North Vietnamese, this third world power

---

Below: *The Republic A-10 did yeoman work with its 30mm cannon and Maverick missiles to wipe out Iraqi armor. Allied pilots called it tank "plinking".*

had a very modern, well-equipped air force and powerful armed forces that had been blooded in the Iraq-Iran War. Consider the following air order of battle:

• Over 950 combat aircraft consisting of 200 support aircraft and 750 "shooters". Support aircraft included a home-built Adnan AWACS type aircraft. The Shooters included a myriad of Soviet aircraft: Mig-21,-23, -25, 27, -29; Su-7, -20, -22, -24, -25; Tupolev -16, -22; Chinese H-6, J-7; Czech L-39; and French Mirage F-1.

These aircraft were armed with French AM-39 Exocet antiship missiles; French AS-10, -11, -12, -14, -20, and -30 stand-off munitions and included the ASL-30 laser guided missile. They also had French Beluga bomblet dispensers; Soviet AA-6 and AA-7 air-to-air missiles; plus French Magic and R-530 air-t-air missiles.

• Seven thousand AAA guns.

• Approximately 16,000 SAMs, including SA-2, -3, -6, -7, -8, -9, -13, -14, and -16; Franco-German Roland; In addition, the Iraqis captured several batteries of U.S. Hawk missiles in Kuwait.

• Modern airfields with extensive shelter protection, consisting of 24 main operating bases and 30 dispersal bases.

• Excellent command and control and communications.

• An excellent in-depth air defense system with layered defenses, centralized headquarters, sector centers, and numerous radar, SAM, flak, and fighter stations.

• One hundred-sixty armed helicopters including the Soviet Mi-24 Hind.[31]

The air defense system was tied

together by a French mainframe computer system built by Thompson-CSF called Kari (Iraq spelt backwards in French). This system was a state-of-the-art air defense system for a third world power based on the best 1970 technology available and was installed in 1988.[32]

Kari had a pyramid type structure. More than 400 observation posts were at the base of the pyramid. Also at the base were 73 radar reporting systems. These observation and radar posts sent their data semi-automatically to the 17 Intercept Operations Centers (IOCs). In hardened concrete facilities, a touch of a light pen could automatically scramble fighters, and supply targeting information to AAA or SAMs. Although in hardened shelters, the IOCs were also in mobile vans inside and could deploy on a moment's notice if attack was imminent. These IOCs reported to four Sector Centers. Thompson computer technology provided simplified systems that could be operated by Iraqi operators with only sixth grade educations if need be.[33]

The Sector Centers fed the Air Defense Headquarters in Baghdad, where a national air defense picture could be assembled. Data was also relayed to terminals in the presidential bunker, intelligence headquarters, defense headquarters, and a Hotel used by the Iraqi leadership. Saddam Hussein could sit behind his 1,800 AAA guns and 60 SAMs surrounding Baghdad, and direct his air defense as he saw fit.[34]

The entire system was tied together by underground redundant land lines, micro-wave relay, and field radios and telephones. Backup communications were provided again and again. All of this equipment was of modern western design.[35]

One of Iraq's unusual systems was a Japanese RM-385 system that could automatically track electronic emissions. Transmissions could be immediately pinpointed for jamming or countermeasures.[36] This system worked well against aircraft and may have helped the Iraqis break up SAR efforts when downed airmen used their unprotected frequencies on their PRC-90 survival radios.[37]

The U.S. Navy had an intelligence unit that zeroed in on the Kari System. By contacting Thompson, the Americans were able to glean a great deal of information without ruining Thompson's chances of selling Kari to other third world countries. During Desert Shield, the Coalition was able to take the measure of the system by progressive ferret and intrusion missions, measuring emissions and responses to various threats. The system was excellent to meet attacks of 20-40 aircraft, but was swamped by the massive attacks planned by the Coalition. French authorities indicated the saturation point was about 120 sorties. The Navy intelligence unit codename SPEAR did fabulous work. Kari was disrupted and ineffective during the war largely due to their efforts.[38]

In theory, the Coalition air forces could avoid most of the Kari defenses by flying low and delaying radar detection. This would also tend to protect any adversary from the many medium-altitude SAMs in the Iraqi system. The Iraqis, well aware of this weakness, tried to seal off the gaps with all types of AAA and low-level SAM weapons. Hundreds of light flak weapons were purchased (about 400 Oerlikons from Sweden alone). Soviet SA-7, -9, -13, and SA-16 IR missiles and along with the radar-guided SA-6 and SA-8 were extensively distributed throughout the Iraqi military. The Iraqis also bought about 300 French

Left: *The U.S. Navy's primary attack aircraft in the Gulf was the F/A-18.*

Roland missiles to augment their Soviet systems. To these systems they could add 5-10 Amnon systems captured in Kuwait. This system combined American Sea Sparrow missiles with an Italian radar and Orelikon guns. Add the Hawks captured in Kuwait and a small number of American shoulder-fired "Stingers" and the low altitude capability of defended targets appeared to be awesome.[39]

The Iraqi army also used the Soviet barrage AAA doctrine that we saw so effective in Vietnam. AAA barrages concentrated on likely targets

and penetration points can be very effective. If gunners are alert, speed is no defense against this kind of fire.[40]

Armed with detailed information about the defense system, the USAF/Coalition planners set out to destroy it and outlined a plan to do so. They named it SEAD for suppression of enemy air defenses. SEAD had five objectives:

• Destroy and disrupt C2 (command & control) nodes

• Disrupt EW/GCI Coverages and Communications

• Force Air Defense Assets Into Autonomous Modes

• Use Expendable Drones For Deception

• Employ Maximum Available Harm Shooters[41]

Using the Israeli attack in the Bekaa Valley in 1983 as a model, the USAF/Coalition idea was to launch a number of drones simulating aircraft into Iraqi airspace. When the radars came up to track the plethora of targets, F-4Gs, EA-6Bs, A-6Es, and FA-18s would unleash the Shrike and Harm missiles they all carried. The Harms unlike the Shrike did not go ballistic if the target radar shut-down. With programmed memory the Harm tracked to the radar's last transmission. Nor did the Harm require recognizable maneuvers for launch like the Shrike and tracking radars had a much harder time determining when the Harm was launched.

FRIENDLY AIR FORCES

Arrayed against the Iraqis was the most staggering air armada in history in terms of capability. Consider:

Coalition Strength

Country	Fighter/Attack	Tanker/Airlift	Other	Total
USA	1323	460	207	1990
Saudi Arabia	276	53	10	339
Great Britain	57	12	4	73
France	44	15	7	66
Kuwait	40	3	–	43
Canada	26	–	2	28
Baharin	24	–	–	24
Qatar	20	–	–	20
UAE	20	–	–	20
Italy	8	–	–	8
New Zealand	–	3	–	3
Totals	1838	546	230	2614

Left: *The A-6 Intruder remained the U.S. Navy's best all weather attack aircraft in the Gulf.*

In pure numbers, Iraq was out numbered in aircraft almost three to one, in shooters 2.45 to 1. In quality and weapons capability the edge must have been four or five to one. In overall training, the gap was wider. Saddam did not have a chance.

Quickly take a look at a compari son of the air to air fighters on the eve of the battle:

Fighter Lineup (Air to Air)
• USAF McDonnell Douglas F-15C, Two Pratt-Whitney F100-PW-100s, 14,870 #dry, 23,830#AB, Mach 2.54 Four Aim-7s, 20 MM, Gatling Gun, Aim-9s.[43]

• USN Grumman F-14, Two GE, F110-GE-400, 14000 #dry, 23,100 #AB, Mach 2.34, Four Aim-7s, Aim-9s, Aim-54 Phoenix. 20 MM Gatling Gun[44]

• RAF Panavia Tornado, Two Turbo-Union RB-199-34R MK 103, 9656 #Dry, 16920 #AB, Mach 2.2, Four Skyflash, Aim-9s, AAM. [45]

• Saudi McDonnell Douglas F-15C, Two Pratt-Whitney F100-PW-100s, 14,870 #dry, 23,830#AB, Mach 2.54

Four Aim-7s, 20 MM Gatling Gun, Aim-9s.[46]

• French Dassault F1C, Atar 9K-50, 15,873 #AB, Mach 2.2, two super 530 AAMs, 2 X 30 MM cannon.[47]

• Iraqi Mig-29, two Klimov RD-33, 11,110 #Dry, 18,298 #AB, Mach 2.3, two radar guided R-27 missile, four R-27 IR missiles (or vice versa),

1 x 30 MM cannon.[48]

The attack opened on January 17, 1991 with unprecedented accuracy and violence. Cruise missiles and Stealth bombers attacked the Iraqi Command and Control system with pinpoint accuracy. New, secret weapons were used against Iraqi power plants and electrical grids and the electrical power system

promptly collapsed. Allied electronic warfare blinded what was left of the radar and communications sites to make way for the strike aircraft to follow. Stealth aircraft knocked out key facilities without being detected. The massive strikes that followed accomplished their missions against minimal opposition. The modern Iraqi air defense system collapsed never to regain its capability throughout the war.

The surrealistic nature of this war was abetted by live broadcasts of air strikes in Baghdad by CNN News. The strike on the Baghdad Telephone Exchange the first night of the war was measured by Allied Air Headquarters when CNN correspondents, who were broadcasting by telephone, went off the air.[49]

The real key to the air battle was the pinpoint accuracy of the Coalition's new "smart" weapons and the effectiveness of the new weapons systems. Even aircraft using "dumb" bombs had unprecedented accuracy from new bombing computers and bombing systems. Civilian casualties were minimized and bomb damage to targets maximized.[50] Allied fighters used modern air-to-air missiles to sweep the skies clean of any Iraqi aircraft that flew. Heavy laser-guided bombs ferreted out the aircraft in hardened shelters. Evidently Saddam thought the sheltered aircraft were "safe" as few planes took to the air. This proved a futile hope.[51]

Over 700 Allied aircraft struck that first night with more than 400 in the first wave. It began with three special assault teams flying into the black, moonless night to clear a way for Pave-Low/Apache helicopter strikes against two Iraqi early warning radars. This Pave-Low mission was to keep a force of deep-penetrating F-15Es from being detected as they ingressed to the target, hugging the ground in the black night while navigating by new modern sensors. Also ingressing through this early warning hole were EF-111s that provided background jamming for the first wave of F-117 stealth birds.[52]

Coordinated with these efforts to the second, ten F-117 stealth aircraft, and 106 ship-launched TLAM Cruise missiles fired by U.S. Navy battleships and submarines struck Baghdad. All these assault forces were designed to clear a path for the strike aircraft to follow. [53]

Although the stealth aircraft were not yet near Baghdad when the EF-111s started jamming a few minutes after the Pave-Low Apaches struck their targets, Baghdad AAA erupted in the now famous barrage dutifully reported by CNN.[54] The defense was blind. Not even one stealth aircraft was hit during the entire war. If the radar operator turned his radar down to escape the jamming strobes, they could not paint the F-117s. The first

F-117s struck Air Defense Headquarters, Intercept Operations Centers, the ATT building and other communications nodes., and approaches to one of the palaces. Baath party headquarters, the electrical power plants, Scud missile production centers, and the Taji missile production center were hit by a number of Tomahawk cruise missiles. F-117s had effective hits on 13 of 17 targets.[55]

As the first F-117s exited the target area, U.S. Navy A-6 fired numbers of TALDs (unguided decoys). Air force BQM-74 drones were launched simultaneously. As the Iraqi defense system reacted to this unending stream of seeming intruders with AAA and SAMs, F-4Gs, EA-6, A-7s, A-6s, and FA-18s filled the air with Harms. The defense was decimated. The F-4Gs alone fired over 100 Harms.[56]

IFF

As we discussed in the Vietnam section, one of the major problems in any air offensive is Identification Friend or Foe (IFF). In the Persian Gulf, the USAF F-15C had two IFF systems. It had the capability to identify Russian aircraft using a classified doppler system and it could interrogate a target's transponder for a coded response. The Coalition rules of engagement required two independent identifications before firing at an aircraft not involved in a hostile act. As the U.S. Navy aircraft had only one identity system, they complained bitterly about these rules. If the aircraft had not been identified

visually, the Navy required AWACs approval to fire. The Navy only had three air-to-air kills during the entire war; however, there were no losses to friendly fire either. IFF was not the problem in this war that it had been in Vietnam.[57]

Two more waves of F-117s followed before the Iraqis could catch their breath. The weather closed in and the stealth fighters were not as effective with their laser-guided bombs as before. Meanwhile non-stealthy aircraft were working over targets all over Iraq. The Iraqi air force would never recover. They lost eight of their best fighters the first night (five Mig-29s, three Mirage F-1s)in the air. Fighter hardened shelters were a prime target for laser and optically guided bombs and missiles. Out of 700 sorties, the U.S. lost only one FA-18 to unknown causes (the Iraqi's never claimed a victory).[58]

It has not generally been recognized or accepted, but in my view, Iraq was pretty well done by dawn on the 17th. Centralized electrical power went out minutes after H-hour (0300 Baghdad time) and did not return until days after the cease fire. The command and control and air defense systems were in shambles. Within days, messages from Baghdad to units in Kuwait were forced to be sent by couriers which took two days. No centralized control was possible for Iraqi Army and Air Force units. They were hunkered down in fortified positions to await their fate.[59] That fate had been assured. The greatest air armada in history was free to range at will and at almost no cost to the Allied Coalition. The dire forecasts of the so-called experts were all wrong. There were very few aircraft losses and very few Allied casualties.

Leaving the initiative to the Allies, Iraq had sacrificed its Air Defense and Command and Control capabilities to the ultra-modern weapons of the coalition. One can only wonder what would have been the results if a determined Iraqi had struck Saudi airfields pre-emptively which were jammed with aircraft practically wingtip to wingtip, most without hardened facilities. Obviously, the Iraqi regime which was consumed with the sustenance of Saddam's power had not developed an aggressive, air force command that was willing to take the risks needed to prevail in the Persian Gulf. A quick review of the Iraq/Iran War shows the same defensive mentality by the Iraqi air force. Saddam does not appear to develop aggressive, independent subordinates. Timid Iraqi air leadership was doomed in the Gulf War.

Allied air was to range over Iraq and Kuwait for 38 days until the ground attack began February 24. During that time, the Iraqi Army's will to fight was largely destroyed by incessant air attack and bombardment.

"The Mother Of All Battles"

When the Allied ground attack began, the Iraqi army was "blind", having no central command and control capability, limited communications, and resupply was being interdicted during both night and day. Individual Iraqi tanks were effectively attacked from the air even when dug-in at night with missiles and 30MM cannon using infra-red and laser sensors. F-111, F-15E and A-10 aircraft discovered that Iraqi tanks could be detected by infra-red sensors at night. The tanks warmed by the sun during the day radiated that heat at night. This heat radiation was exacerbated by their tank engines if they had been run during any previous time period. The fighter jocks could clearly see the tanks through their IR sensors against the colder desert. It was almost like a

Below: *The Vought A-7E was being replaced by the F/A-18 when Desert Storm occurred. A-7 "Slufs" were still in fleet service on some carriers and flew missions in the Gulf.*

Above: *During the Gulf, the majority of USAF attack sorties were flown by F-16s. When the bombs were gone the F-16 remained a formidable air-to-air opponent.*

deadly computer game. The aircrews called it "tank plinking". The laser guided bombs and IR Maverick missiles were particularly deadly.

The Iraqi army was totally demoralized, it had been severely mauled, and was unable to maneuver without being pounded from the air. Its deterioration was nearly complete. Many units were down to 50 % manning. Air Power had opened the door.

The Iraqi Army's attack at Khafji made the case. This was a large well-planned attack by three heavy divisions and elements of two Corps. The night before the attack, a Marine outpost spotted two large armored columns. The outpost called in air strikes that devastated the Iraqi Brigade designated to lead the attack and it continued throughout the battle. The airpower committed, destroyed the majority of two divisions, caused over 2000

Below: *The Patroit Missile System is the only defense the free world had against the Scuds of Iraq. It is still the only operational anti-missile system in existance. The Israelis are working very hard to perfect a follow-on system called "The Arrow" System.*

Patriot

casualties and the loss of 300 vehicles. When the Iraqis left their holes they were decimated and were paralyzed by air assault. An aura of hopelessness permeated Saddam Hussein's army following this attack.[60]

Schwarzkopf built up his ground forces until he had an effective offensive force. Using his amphibious forces as a feint in Kuwait, he launched a massive envelopment of Saddams' right flank, using the French and American airborne forces to block communications and retreat routes toward Baghdad and his heavy armor in a crushing main attack closer in on the left flank. A secondary attack was launched by Marines and Coalition Arabs directly into Kuwait to fix Iraqi forces there. All ground attacks were supported by overpowering close-support and interdiction by both tactical fighters and helicopters. Every ground attack was successful beyond all expectations. It was all over in 100 hours. The only organized resistance came from isolated elements of the renown "Republican Guard" who proved to be minor leaguers in combat with the well-equipped Allies. Iraqis everywhere surrendered in droves. U.S. Army casualties totaled 366 many of whom were victims of Allied "friendly fire". Iraqi casualties were in the thousands with over 00,000 taken prisoner.[61]

War On The First Team

US Army equipment proved to be absolutely superior. M-1A tanks

were devastating to the Russian equipped Iraqis. Even the new T-72s were ineffective against the M-1As. On a number of occasions the M-1A was hit first and uninjured, then knocked out the enemy tank. The missile equipped Bradley personnel carriers also were very effective. Even its 25MM Chain Gun proved capable against light Iraqi vehicles and light tanks. With Allied command and control, missile technology, control of the air and unprecedented firepower, it was all over before it started. Saddam's battle-hardened veterans were incapable of playing in the major leagues. They were not first-team capable. Critics and so-called experts who degraded American equipment and capabilities were proved dead wrong. There were many red-faces among the military "experts" who had been full of doom and gloom prior to the war.

The Scuds

Saddam did deploy one weapon system for which we had no ready answer, the Scud missile. Air commanders using erroneous intelligence, had assured General Schwarzkopf that they would take care of the Scuds. They proved to be wrong. Intelligence had assured Air Force leaders that it would take several hours to launch the missiles and that there were very few mobile launchers. Both estimates were grossly incorrect.[62]

Nor was the USAF able to completely disrupt Saddam's radio communications with the Scuds in western Iraq.[63]

The Scud was employed by Saddam as a political/terror weapon against

Israel and Saudi Arabia. Although the weapon system has nuclear, chemical and biological capabilities, so far as can be determined it was used only with conventional warheads. Used this way, the missile had the characteristics of a longer range German V-2.[64] With no precision guidance, the Scud was purely an area weapon. The randomness of this area missile almost seemed to add to its terror effect on target populations.

Saddam wanted to get Israel in the war which he thought would break up the Allied Coalition. Fortunately, because of Israeli courage and restraint and pressure from the U.S., this did not occur.[65]

The U.S. Patriot missile system proved to be at least partially effective in intercepting the inbound Scuds. Unfortunately, interception did not prevent portions of the missile and debris from impacting somewhere in the target area as the Patriots blew the Scuds apart.

An inordinate amount of effort went into detecting Scud launchers and sites with limited success. A 1950 missile and its mobile launchers proved all but impossible to locate and eliminate. Airborne patrols, road-watch teams, sophisticated sensors in space, etc. were employed with mixed and never satisfactory results.

The employment of the Scud against Israel was the major factor driving the anti-scud effort.

The frail Allied coalition with the Arab states (Saudi Arabia, Egypt, & Syria) could have been destroyed if Israel had joined the Allies. With their populations pro-Saddam, Israel joining the coalition or unilaterally attacking Iraq probably would have destroyed all of President Bush's exhaustive diplomacy. The leadership of the U.S. and Israel was solely tried by these Scud attacks. The U.S. promised Israeli leaders that the Allies were doing everything possible to eliminate the Scuds. This took some convincing to the always aggressive Israeli leaders, but prudence prevailed and the Israelis stayed out of the war. Hence the extensive search and destroy efforts to locate and destroy the scud launchers.[66]

EPILOGUE AND THE FUTURE

The Persian Gulf War was one of the most successful in U.S. history. It was fought with clearly defined, attainable objectives and U.S. leaders stuck to their objectives. It was also the first war in history in which airpower, not ground forces played the dominant role.[67] To defeat a nation with one of the largest and most modern military establishments in the world with less than 400 casualties was nothing short of a miracle. It clearly proves the author's premise that casualties are inversely proportional to the predominance of force committed by the stronger foe. It illustrates the idiocy of Maxwell Taylor's "Flexible Response" once and for all. Mass and Economy of Force are sound principles of war. The Flexible Response ideas of Vietnam days are contraventions of these principles. One ignores the lessons of history about these two principles, only at national peril. Whatever history says about George Bush, he was the first U.S. wartime President to understand this since WW II. The

results speak for themselves.

In their books on the war both Bradford and Dido (see Bibliography) question if war was really necessary. Bradford makes this point upfront as he used the title, *George Bush's War*. I believe the war could have been avoided, but not with George Bush as U.S. President after the invasion of Kuwait. It was not in the U.S.'s or the world's interest to have Saddam Hussein controlling 70% of the world's oil.

Saddam might have gotten away with taking the long-disputed Persian Gulf islands and the Ramilia oil fields. Ambassador Glaspie indicated as much in their July meeting when she intimated that the U.S. was not concerned with "middle-eastern border disputes" and Ramilia and the off-shore islands had been in border dispute for decades. It was the shock of overt Iraqi aggression, the invasion of all of Kuwait, and the adamant refusal to back off that spurred President Bush into action. With the invasion, the U.S. was genuinely concerned about Saddam's threat to Saudi Arabia. Three U. S. President's had plainly and stoutly proclaimed the Middle East oilfields as a key area of U.S. interest.

It is difficult to imagine that Saddam realized the Pandora's box his attack on Kuwait would open. Once he decided to invade Kuwait however, he would have been better served to have attacked Saudi Arabia to close her air bases and seaports to resupply and reinforcements from the West. Iraq then could then have defeated the Coalition forces in-detail as they arrived, if they tried to resupply the Saudis. If the Saudi ports had been interdicted with mines, submarines, and air power; the coalition would have had a diffi-cult time supporting a Desert Storm type offensive. Neither Saddam's moral courage or Iraqi military power proved equal to this challenge. The Iraqi Air force, in particular, proved incompetent and even cowardly. They simply did not fight well.

It also must be realized that Saddam had to worry about the U.S. nuclear arsenal. Americans often forget we have this capability. Pushed with a fait accompli such as the loss of Saudi Arabia and control of 70% of the world's oil would the U.S. consider nuclear options? It is an interesting question to us, but had to be a deadly serious one to Saddam Hussein. We are the only nation that has up to now used "nukes".

Left: *The heralded conflict between the Mig-29 Fulcrum (pictured) and the F-15 Eagle did not occur. F-15 BVR missiles destroyed several Mig-29s with no retaliation. The message may be the **best missile system** wins not the best aircraft.*

In further mitigation, Saddam's regime had to stay in power by force and the Air Force by its nature had to be tightly controlled. Pilots are free and independent to act when airborne. Such independence is necessary for an aggressive, accomplished air war component, but could be totally deleterious to a tyrant. One cannot have it both ways. Therein may lie the story of the Iraqi Air Force's absence/failure during Desert Storm.

Incidentally, if Saddam had built a modern, attack submarine force, say patterned after Great Britain's fine subs, the Coalition might still be fighting. Rapid reaction shipping was in short supply. A mod-

---

Below: *Barrage electronic jamming was done by EF-111s. The jamming ebvironment aided the stealth employed by F-117s.*

ern nuclear sub fleet could really exploit this U.S. weakness. Nor are carrier strike forces safe in proximity to such threats. Fortunately, this technology was not yet available to the Iraqis in 1990. This lesson was not lost on Iran, India, and Communist China who are today buying Soviet submarine technology on a crash-priority basis. At their present rate, all three will be attack submarine powers by the end of the decade. This build-up must be carefully monitored by U.S. intelligence. A Persian Gulf type-emergency in the year 2000 might be a different kettle of fish for U.S. strategic planners. The next Colonel Warden will in all probability contend with some new "centers of gravity".

The most domineering aspect of this war was the pervasiveness of Coalition Air Power. Fanatical Air Power adherents characteristically proclaimed this as the first war won by air alone. Not quite so, but surely they had more justification this time

than ever before. General George Kenney in WW II, used to say he wanted MacArthur's doughboys committed to amphibious attacks to be able to "walk ashore with their rifles slung on their backs" because of his air preparation. That was the aim of Coalition Air in the Gulf, and like Kenney, they came close. Coupled with overall Coalition superiority in material, strategy, and command – Air Power lead by fighter aviation had destroyed the Iraqi Army's will to fight. It was still necessary to occupy the ground to impose our will, but occupying that ground was easier than ever before, largely because of firepower delivered from the air both prior to, and during, the land battles.

Of particular importance in the Persian Gulf, was the incredible accuracy and power of air delivered weapons. "Surgical strikes" had been long touted, but only now delivered while the whole war watched on TV (the first employ-

ments were in Vietnam and got little or no public exposure). The impact of LASER/Optically guided munitions was driven home in living rooms around the world by television sensors that showed guided munitions striking pinpoint targets with devastating results. A well-orchestrated management of overall war news emphasized these graphic results. Nor could the results of cruise missiles, and stealth aircraft in destroying heavily defended targets be missed. Their would be no revulsion of the war effects by the folks back home this time. They cheered the Coalition to victory.

The Coalition attack on the Command and Control "center of gravity" as defined by Colonel Warden was decisive. Its destruction led to the collapse of the Iraqi air defense system. With the defense system gone, the Coalition air ranged the skies for over seven weeks, literally attacking at will every other "center of gravity" in Iraq, including its army.

No answer was found to the problem of mobile ground-to-ground missiles. Coalition air power and road watch teams inserted into Iraq were not successful in stopping the Scud threat. The "Patriots" provided our only shield from these short-ranged, random firings. If these missiles had been equipped with nuclear warheads, the results could have been horrendous. Even with biological or chemical warheads they would have been frightful. With mobile missiles of intercontinental range a reality in the Russian Republics, and their being sold to Iran, India, and China today, U.S. strategists have a myriad of problems to face in the future. The fanatical nature of our prospective enemies in the future heightens the concerns. In light of these facts, the stopping of the Strategic Defense Initiative (SDI) appears ill-advised and perhaps really foolish. An answer is yet to be found, but one must be found.

NOTES

1. Smith, Jean Edward; *George Bush's War*; Henry Holt Books; New York; 1992; pp. 26-28.

2. Op. Cit.; pp.21.

3. Ibid.

4. Op. Cit.; pp.31.

5. Gardner, Gerald; *The Shining Moments, The Words and Moods Of John F. Kennedy*; Pocket Books Inc.; New York; 1964.

6. Op. Cit.; pp. 53-58.

7. Op. Cit.;pp. 16.

8. Op.Cit; pp.23.

9. Hiro, Dilip; *Desert Shield To Desert Storm - The Second Gulf War*; Routledge; New York; 1992; pp. 102-110.

10. Atkinson, Rick; Crusade, *The Untold Story of the Persian Gulf War*; Houghton Mifflin Comapny; Boston; 1993; pp. 516.

11. Hiro; Op. Cit.; pp. 316.

12.	Smith; Op. Cit.; pp 17-19.

13.	Op. Cit.; pp.66-67.

14.	Hiro, Dilip; *Desert Shield To Desert Storm*; Routledge; New York 1992; pp. 114-116.

15.	Summers, Harry G. Jr.; *On Strategy II*; Dell Books; New York; 1992; pp. 190-191.

16.	Hiro; Op. Cit.; pp 316.

17.	Summers; Op. Cit.; pp. 44-57, & pp. 181-182.

18.	Hiro, Dilip; *Desert Shield-Desert Storm The Second Gulf War*; Routledge; New York; 1992; pp.340.

19.	Op. Cit.; pp 313.

20.	The author had two sons-in-law who served in Desert Shield in the 82nd Airborne Division. They still joke about their fears as a "Speed-Bump" for Saddam.

21.	Freedman, Lawrence and Karsh, Efraim; *The Gulf Conflict 1990- 1991*; Princeton University Press; Princeton New Jersey; 1993; pp. 88.

22.	Gordon and Trainor; *The General's War*; Little, Brown & Company; Boston; 1995; pp. 234.

23.	Freedman and Karsch; Op. Cit.; pp. 204.

24.	Freedman and Karsh; Op. Cit.; pp. 204-210. 25. Hallion, Richard P.; *Storm Over Iraq*; Smithsonian Institution Press; Washington,D.C.; 1992.

26.	Op.Cit.; pp 162.

27.	Op. Cit.; pp 1-89.

28.	Atkinson, Rick; *Crusade, The Untold Story of the Persian Gulf War*; Houghton Mifflin Company; New York; 1993; pp 57-65.

29.	Hallion, Richard P.; *Storm Over Iraq*; Smithsonian Institution Press; Washington, DC.; 1992; pp. 152-161.

30.	Ibid.

31.	Op. Cit.; pp. 146-147.

32.	Gordon, Michael R. and Trainor, Bernard E.; *The General's War*; Little, Brown, And Company; Boston, MA.; 1995; pp. 111-116.

33.	Ibid.

34.	Ibid.

35.	Ibid.

36.	Ibid.

37.	Op. Cit.; pp. 274-275.

38.	Ibid.

39.	Op. Cit.; pp. 115.

40.	Ibid.

41.	Op. Cit.; pp. 117.

42.	Hallion, Op. Cit.; p. 158

43.	Green & Swanborough; Op. Cit.; p.371.

44.	Op. Cit.; pp. 269-271.

45.   Op. Cit.; pp.462-463.

46.   Op. Cit; pp. 371.

47.   Op. Cit.; pp. 155-157.

48.   Op. Cit.; pp. 402-405.

49.   Hallion, Richard; *Storm Over Iraq, Airpower and the Gulf War*; Smithsonian Institution Press; Washington, D.C.; 1992; pp. 171

50.   Hallion; Op. Cit.; Appendix B; pp. 282-288.

51.   Hallion; Op. Cit.; Appendix E; pp.303-307.

52.   Gordon & Trainor; Op. Cit. pp. 219-227.

53.   Ibid.

54.   Hallion; Op, Cit.; pp 166-177.

55.   Op. Cit.; pp. 225-229.

56.   Ibid.

57.   Ibid.

58.   Op. Cit.; pp. 234-235.

59.   Ibid.

60.   Ibid.

61.   Hallion; Op. Cit.; Table 7-2; pp. 238.

62.   Gordon & Trainor; Op. Cit.; pp. 239-243.

63.   Op. Cit.; pp. 235.

64.   Atkinson; Op. Cit.; pp. 145.

65.   Op.Cit.; pp. 240.

66.   Atkinson; Op. Cit.; pp. 144.

67.   Gordon & Trainor; Op. Cit.; pp. Preface xi.

Above: *A bevy of "relaxed stability" fighters are being developed to out perform the F-14, F/A-18, F-16 and F-15./ The French Dassault Rafale is pictured here.*

CHAPTER 13

THE FUTURE

INTRODUCTION

Recognizing the importance of air superiority in any potential conflict, every military power in the world today is trying to develop a new, viable, next generation fighter, i.e.:

- U.S. – the Lockheed F-22 Lightning II

- Britain/Germany/Spain/Italy – the Eurofighter

- France – the Rafale

- Sweden – the Gripen

- Russia -- the Sukhoi-35, the Yak-141, and others

- China -- all of the Russian developments that can be purchased with or without Chinese modifications

In addition, present fighter (F-15, F-16, Mig-29, Su-27, Mirage 2000, etc.) configurations have been, and are being, modified to improve aerodynamic performance until these next generation fighter developments can be brought to operational status. The target aircraft are the American "Teeners" F-14, F15, F-16, and FA-18s which have dominated air battles since the early 1980s. These upgrades are taking many different forms:

- upgraded and improved engines

- straight wings with chines

- delta wings with canards

- variable-sweep wings

- cranked arrow wings

- forward swept wing with canards [1]

Coupled with these aerodynamic improvements are weapon systems changes: better AAM missiles, improved ECM gear, improved ARMs, AGMs, and smart bombs. The race is on.

DESIRABLE CHARACTERISTICS FOR THE FUTURE

INTRODUCTION

We have watched the fighter evolve since WW I into a missile-carrying supersonic platform. The technological trend of our time is the missile, the SAM, the AGM, and the AAM. Equally important are electronic warfare weapons to overcome these missiles. Present U.S. plans do not appear coherent in this latter area as both the EF-111 and F-4G Wild Weasel are currently being phased out. It seems as if the budgeteers in their enthusiasm to cut the Defense budget ("the Cold War is over, we no longer need a defense") are going to forget the lessons of Vietnam, Yom Kippur, the Bekaa Valley, and the Persian Gulf War.

Hopefully, our next generation fighters will be ECM self-sufficient and carry their own ECM protection from SAMs. However, as I was writing this, Captain Scott O'Grady's USAF F-16C was shot down by a Bosnian Serb SA-6 missile. News stories indicate Captain O'Grady was neither carrying ECM pods or supported by ECM dedicated aircraft. [2]

Did his ECM support get cut by some overzealous budgeteer in Washington? It does not appear to be an auspicious beginning for the post-Desert Storm era.

V/STOL and ASTOVL* capabilities are currently getting a lot of attention by designers. It is not well known, but in 1975 the U.S. Navy conducted some air-to-air tests between the Marine V/STOL Harrier and the F-14 Grumman Tomcat. The results were surprising. With a good missile system, the Harrier often prevailed. The V/STOL characteristics of the Harrier introduced a number of new parameters into the air-to-air equation. The U.S. Navy, for one, was not surprised by the Harrier's performance in the Falklands. Vectored thrust and V/STOL capabilities appear to have valid application in tomorrow's fighter operations (see the Falkland discussion in Chapter 10)."

*NOTE:
V/STOL = Vertical/Short Takeoff and Landing
ASTVOL = Advanced Short Takeoff and Vertical Landing
Times are obviously changing for the fighter. Will the new fighter

and missile capabilities upset the principles that we have developed through the previous chapters? Only time will tell, but perhaps we can get a glimmer by examining what we have previously determined from our studies. From a careful re-study of what we have learned from the past, it does appear we might determine some desirable characteristics for the next generation of fighters.

For example:

• Super-sensitive sensors to provide beyond visual range ID

• "Smart", accurate, effective weapons, missiles, guns, and bombs

• Speed and energy

• Extraordinary maneuverability

• High power-to-weight ratios (ability to maneuver without loss of energy)

• Vectored thrust, V/STOL, ASTVOL capabilities*

• Stealth properties

• Self-contained ECM

Let's discuss each characteristic one at a time:

SUPER-SENSITIVE SENSORS

Being able to tell the bad guys from the good guys is imperative in a world where the air war will primarily be fought beyond visual range. The fighter hopefully will have its own systems to do this, but

airborne radar (AWACS) will continue to play a role along with Global Positioning Satellites (GPS). Current F-15s can interrogate transponders and identify certain engine characteristics using sensors and on-board computers. During Desert Storm, with these systems and AWACS, visual identification was not the requirement we had in Vietnam. Long-range radar missiles were employed with efficiency. Look for new developments in these air-to-air areas. The trend is for the side with the best and longest range

AAMs to gain air superiority (Falkland Islands, Bekaa Valley, Desert Storm). We must push development in this area. As I have already indicated, we probably have the second best IR missiles today. BVR ID is critical to the modern "fire and forget" missiles we need. Using fire and forget missiles, a modern fighter can lock on to a target, fire and see it off (the missile destroys this original target), while breaking-off to defend or attack another set of adversaries.

mance of the AIM-9M versus the AA-11 Archer and the costs of the AIM-9X. These proponents feel the turning close-in IR missiles may be overrated. If the AMRAAM can reach BVR and eliminate the threat, why spend the money for a complex, expensive IR turning missile?[4]

The Israelis, aware of the AA-11 capabilities, characteristically have not deferred to this argument, nor have they waited for the U.S. to commit to the AIM-9X. With their small country, close-in combat is a way of life for the Israelis. They have gone operational with the French Rafael Python said to exceed the proven capabilities of the AA-11. Unfortunately, the Python does not readily match with U.S. weapons systems according to U.S. officials. However the Israelis went operational with Pythons in 1994. The Israelis use largely American aircraft.[5]

The Germans have bought 75 Mig-29s equipped with AA-11s. USAF F-16 have exercised with these German Migs and the AA-11s have given them a marked advantage although the F-16s can outmaneuver the Migs.[6] The Russians still insist the AA-11 is the best IR missile in the world and it certainly outperforms the AIM-9M. For the first time since early WW II American pilots are flying clearly inferior equipment. The future could make this disparity worst.

AIR-TO-AIR WEAPONS

Missile systems with beyond visual range (BVR) and fire and forget capabilities are a key to the future. The IR family of missiles (Aim-9L & M, AA-11) over the years have been the most effective AAMs as they were "fire and forget". With the introduction of the AIM-120 AMRAAM radar missiles at the end of the Persian Gulf War, the radar guided missiles have entered this arena of effectiveness. In general, these radar-guided missiles usually have a longer range than their IR contemporaries and of course are effective in weather. We can expect more of this family in the future. Both radar and IR missiles will have enhanced performance in close-in, hard-turning fights and the missiles will provide the final guidance to target themselves.

There is a real debate going on between the adherents of the AMRAAM and the shorter range IR systems. This debate has been precipitated by the now inferior perfor-

215

All of these new systems use helmet directed systems and vectored thrust to enhance missile performance. In effect, if the pilot can see his opponent, the opponent is dead with these systems. Missiles are said to be able to circle to impact a fighter separating in the opposite direction.[7]

Meanwhile, the Russians are improving the AA-11 Archer[8] and the U.S. and Britain are having a fly-off test between the AIM-9X and the British ASRAAM.[9] Only time will tell the outcome. However, if the U.S. fights an opponent equipped with AA-11s or Pythons in the near-term (before a new missile is operational), we will be at a marked disadvantage in air-to-air combat.

The gun will become less important in air-to-air combat between conventional aircraft, but will still be necessary for fill-the-windscreen close-in combats. It has generally been overlooked, but the gun could also become very important between stealthy opponents, particularly when the opponent is so stealthy that radar and IR guidance is limited. The M1-A1 eyeball still works in this environment to guide the bullets if you can see the opponent.

Interestingly, laser systems designed to attack the eyeballs are rumored to be in the works. If true, we can expect a number of blind casualties among the aircrews of the future. "Blind Flying" could take on a new meaning.

AIR-TO-GROUND WEAPONS

Air-to-ground weapons have come of age with the smart family of bombs and fire and forget missiles. The IR guided missiles (e.g., the Maverick) really came into their own during Desert Storm when it was realized that armored vehicles heated up during the day and became standout targets against the cold desert during the hours after dark. This enabled the Maverick equipped aircraft to "tank-plink" all night. Laser guided bombs were obviously effective against Iraqi hardened targets as the American public can attest from video tape released to public networks. These smart weapons will get better as time provides the opportunity to upgrade and improve them.

Not only are smart bombs still improving in accuracy, but dumb bomb accuracy is getting better also. This is due to better aircraft systems. With Global Positioning Satellites (GPS) to update aircraft inertial system accuracy, with better computers, and helmet aimed systems, we can expect accuracy for guns, dumb bombs, and rockets

Right: *The Eurofighter is being built by a European Consortium to replace their present fighters and to exceed the performance of the Mig-29, Su-30, F-16 and F-15.*

to improve markedly. Automatic bombing systems are approaching laser-guided weapons in accuracy.

Incidentally, we can expect laser and optically guided countermeasures to be developed as the defensive-offensive cycle reverberates.

SPEED AND ENERGY/THRUST TO WEIGHT RATIOS/MANEUVERABILITY

It is interesting to note that the 1960-1970 fighters (F-4, Mig-21, etc.) had top speeds in the Mach 2.0–2.5 range as do the next generation F-22, Rafale, Euro-fighter, and Su-27/Su35. This speed regime is dictated by present aircraft structure and operating altitudes.

The only exceptions to these speeds were the Mig-25 and SR-71 type aircraft which were in the Mach 3.0-3.5 category. To reach these Mach 3.0 speeds safely, engineers had to go to steel and titanium construction and were forced to operate at very high altitudes in near space. Airframe heating from these high machs forced engineers to use heat resistant materials. This airframe heating made these aircraft the world's best IR targets (one reason the B-70 and high altitude version of the B-1 were abandoned) for IR guided missiles. The SR-71 is used for high altitude reconnaissance (it recently was reactivated for use in Bosnia) and the Mig-25 for non-maneuvering targets (i.e., B-70 types). Both have limited maneuverability in the dense, lower atmosphere.

The follow-on aircraft of the mid-70s (e.g., F-14, F-15, F-16, and FA-18) all had Mach 2.0-2.5 performance, but unprecedented thrust to weight/drag ratios (higher than 1.0). This enabled them to maneuver without loss of speed or energy, and to accelerate under high drag circumstances and even in vertical climbs.

With "relaxed stability" controlled by computers (F-16) maneuverability took on quantum improvement. The next generation will enhance this capability with vectored thrust and the ability to maneuver at high angles of attack beyond stall. Forward mounted canards, improved computers, fly-by-wire, and swept forward wings will improve this even more.

Engine improvements will provide thrust to weight ratios sufficient for supersonic flight without afterburning which will markedly improve fuel consumption.

Speed may make a major comeback if someone develops a fighter that can transition from an aerodynamic fighter to a space fighter (orbital) and back to aerodynamic. This will require a major revolution in powerplants, but the science is there and it might be done in the near future. Such a breakthrough

would revolutionize *aerospace* superiority overnight.

VECTORED THRUST-V/STOL-ASTOVL

The lessons of the Harrier in the Falklands are still fresh in everyone's mind. The British Harriers armed with Aim-9Ls gained air superiority over the Argentine Air Forces and were able to operate in the harsh South Atlantic weather from carriers when conventional carrier operations were impossible.[10]

The Harriers proved enormously flexible. Any ship with a helicopter deck was a landing field as well as any open space, road, or short runway. The ability to operate and land in low clouds and visibility was particularly useful in the abominably bad weather of the Falklands. It is almost certain that conventional fighters would have sustained severe operational limits and losses in these conditions and some operations safely conducted by Harriers would not have been possible with conventional carrier aircraft.[11]

In addition, the Harriers proved serviceable and quite robust under extremely trying operational conditions. Sortie rates were high and extraordinary in-commission rates were sustained.[12]

Prior to the Falklands, the small size of the Harriers was seen as a detriment. Unexpectedly, this proved to be a real advantage in air-to-air engagements as the sea camouflage, small size, and low visibility weather combined to make the British V/STOL fighters difficult for Argentine pilots to see. Significantly, in many of the Argentine aircraft destroyed, the pilots indicated that they never saw the Harrier that shot them down. Some did not see the Harriers until they fired their missiles which proved too late. Conversely, the larger, lighter colored Argentine fighters and attack aircraft stood out against the dark land/seascape.[13] The natural gloom of the Falkland winter appeared to have worked in favor of the British Harriers.

Although Argentine tactics and deployments left much to be desired, the efficacy of the Harriers against their Mirage equipped opponents was lost on no one. Ironically, the French unsuccessfully attempted to design a successful V/STOL Mirage, the Mirage IIIV in the 1960s. Attempts to develop this first supersonic V/STOL aircraft were abandoned after a second prototype crashed in November 1966.[14]

Designers and engineers immediately began searching for ways to incorporate the V/STOL advantages of the Harrier into modern fighters without losing the performance of the fighter. That is much of what is going on today.

VECTORED THRUST

The application of vectored thrust gave the Harrier some slight advantages over the conventional Mirage/Daggers in air combat maneuvering (it should be recognized there was only one real air-to-air engagement in the Falklands where Argentine Mirages tried to "scissors" with the Harrier), particularly when equipped with a superior all-aspect IR missile which made up for Harrier's overall speed deficien-

cies. Since vectored thrust is the easiest of the Harrier's advantages to design into a supersonic aircraft, this was one of the first features to be incorporated (e.g., Eurofighter, F-22, Su-35) in the new designs.

Incidentally, vectored thrust only improves the Harrier's turn capability by about a half "G" of acceleration (turn capability varies directly with G capability up to the stall at a given airspeed). The capabilities of V/STOL and ASTOVL aircraft to rotate the nose at almost any speed and to rapidly decelerate and accelerate however, introduced a whole new ball game to air combat maneuvering.[15]

V/STOL (Vertical and Short Takeoff and Landing)

Besides the Harrier which is used by the British and Americans, the Russians use the V/STOL Yak-36 and Yak-38 on several "jumpjet" carriers which have all of the advantages which we saw in the Falklands with the Harriers.[16]

Advantages

V/STOL aircraft have obvious operational advantages:

• The ability to use short fields for takeoffs and landings

• VTOL operation particularly when landing (when light)

• V/STOL operation off jump-jet carriers

• VTOL operation in near zero-zero visibilities
Disadvantages

V/STOL aircraft do have disadvantages:

•Afterburning (AB) with V/STOL engines is technically complex. Using AB in the vertical plane tears up ground surfaces and ship decks, and creates takeoff turbulence.

• V/STOL aircraft with several engine nozzles, lift-fans, etc. are inherently aerodynamically dirty and therefore, usually subsonic. Flow problems around lift nozzles get very interesting supersonically. The French Mirage III-V and Russian Yak-141 are illustrative. Both of these attempts at supersonic V/STOL designs were discontinued after significant development problems.

• Another V/STOL problem is aircraft payload. As engine thrust must overcome aircraft weight and payload for a vertical takeoff, payload, of necessity, is limited. This restricts the viability of V/STOL aircraft as fighters and fighter-bombers.

To summarize: viable V/STOL aircraft of today are sub-sonic and payload/range is limited. But their short takeoff and landing capability is a significant advantage that cannot be denied.

ASTOVL (Advanced Short Takeoff And Vertical Landing)

The development of ASTOVL aircraft was undertaken in order to overcome the disadvantages of V/STOL aircraft while maintaining the short takeoff and vertical landing capabilities that are so useful.

The idea is to use an afterburning, powerful, light engine to provide short-takeoffs within about 350 feet (within the length of a modern carrier deck) thereby avoiding the payload restrictions associated with vertical takeoff. The lift ducts used for thrust vectoring for takeoff and landing are designed to close for cruise or high speed flight allowing supersonic flight. Then the aircraft can recover vertically once its fuel load and ordnance are expended and therefore the aircraft is much lighter.[17]

In this way, the payload limitations of V/STOL can be avoided, the ASTOVL can be a fully supersonic fighter, and recovered vertically on practically any ship or dispersed field – we can have our cake and eat it too! And the cake has icing as the fans allow thrust vectoring for air combat maneuvering as well.

An 86% size model of the Lockheed ASTOVL fighter began testing in June 1995 and is about three years from possible operations. The prototype looks very much like an STOL version of the F-22. Much of the aerodynamic data from the F-22 is being used. This allows Lockheed and the engine manufacturers to concentrate on the ASTOVL engine as much as possible. A Pratt-Whitney F-100-220 engine powers both the jet and fan portions of the engine on the Lockheed ASTOVL model.[18]

STEALTH PROPERTIES

Any fighter designed for the future must have stealth characteristics. Non-stealthy designs will be nothing but glorified missile targets in a few years. Radar cross-section (RCS), radar reflectivity, and IR signature must be reduced to the absolute minimum. Very little information is available even on current stealth fighter designs as most is highly classified, but some general characteristics are known: IR signatures are reduced by exhaust mixing that cools exhaust.

Engines are blended into the fuselage/wings with intakes usually mounted on top of the wings to reduce RCS. Tailless, flying wings (e.g. the defunct A-12 and very alive B-2 programs) offer much lower RCS than aircraft with large vertical and horizontal tail surfaces. These general postulates explain the attractiveness of flying wing designs for stealth.[19]

The flying wings, abandoned in the late 1940s because of inherent aerodynamic instability, are useable for stealth today since computer controls are now available to manage this instability. In fact, the instability can enhance maneuverability with the instantaneous, precise control movements possible with modern computers. Composite fiber structure is extensively used for construction to provide strength and stiffness without excessive weight and radar absorbing material (RAM) is also used where appropriate.[20]

Weapons, ECM pods, and fuel tanks are carried internally to reduce radar cross section. This means internal bays must be designed to accommodate stores which have usually been carried externally in the past. ECM pods, which require sophisticated antennae to operate properly, are a challenge to design and carry internally.

ECM SELF-SUFFICIENCY

As indicated above, the next generation fighter is planned to be self-sufficient and supporting against electronic threats. This includes IFF, radar-guided and IR guided missiles, and defense against surface launched missiles. It also involves sensitive sensors to detect and identify aircraft and enemy electronic sensors. High speed computers, using artificial intelligence, must then select the proper countermeasure for the particular threat.

The O'Grady shootdown in Bosnia, already discussed, has raised questions about the USAF possibly buying F-16 aircraft without state-of-the-art ECM protection. No aircraft of the future will survive without an ECM capability against a major enemy. One can infer that those kinds of budget constraints are false economy and may cost lives and possible victories in the future.

SOME NEW CONCEPTS

MMSA

Lockheed is famous for its futuristic designs from its famed "Skunk

Left: *The Su-30 was specifically designed to exceed the performance of the F-15 Eagle and it is in operational service in Russia and versions are being sold to China and India.*

221

Works". From that famed shop has come the idea of a multiple mission support aircraft or MMSA. This concept is very interesting as it is based on a "lift-fan in the wing" flying wing concept which would be exceedingly stealthy, have a large payload capability, and possess V/STOL short-field capabilities. It could replace medium lift helicopters and tilt-rotor machines with a stealthy, faster machine. It could drastically affect fighter operations both in close support and ECM support roles. Larger versions also would make an excellent V/STOL airlifter or commercial feeder liner.[21]

EX: THE DIAMOND WING

The Boeing Company, when asked to consider a replacement for the E-2 carrier based AWACS aircraft, came up with a diamond shaped "joined wing" design where the radar antennae were enclosed in a severely modernized bi-plane design (the main wing and vertical stabilizers are joined in a futuristic bi-wing design that looks like a diamond). This design allows phased array radars antennae to form part of the wing structure and skin, dramatically reducing drag in comparison to present heavy, rotating, aerodynamically dirty overhead antennae. Payloads, range, and loiter capabilities for these aircraft would exceed any other existing carrier AWACS. AWACS aircraft are essential to U.S. Navy command and control of fighters and the EX would represent a marked increase in capability.[22]

"AURORA"

As mentioned earlier, one character-istic that would be extremely useful to a future fighter would be the ability to perform at very high altitude or near-space beyond the operating altitudes of the SR-71 where the atmosphere no longer restricts aircraft performance. The ability to operate in both space and the atmosphere would provide enormous potential. The capability to transition to and from space to the upper atmosphere would also be an exponential improvement in performance. Speeds in space can be incredible. (Orbital velocity is roughly 25,000 MPH). There have been rumors of such aircraft for years, but if they do exist they are so secret there are very few details available.

An inadvertent release of a 1985 program document showed a huge $2.1 billion line item called "Aurora". The item followed the line item for the SR-71 "Blackbird".[23] This fueled the idea that Aurora was a new, space-based reconnaissance vehicle. The New York Times followed with a report of a reconnaissance vehicle that far surpasses the Mach 3 and 85,000 feet capability of the "Blackbird". When the SR-71 was shelved a few years ago, many believed Aurora was operational. This conclusion was based on the history of the U-2 and SR-71. When the U-2 flights over the Soviet Union were suspended by Eisenhower in 1960, we did not lose any reconnaissance capability. Both satellite systems and the SR-71 had reached operational status by the time the U-2 was suspended.

As soon as these reports appeared, seismological recordings which recorded sonic booms in Southern California began pointing to possible hypersonic flight paths and reports of sightings near classified airfields near Groom Lake, Nevada began coming in. An aircraft was sighted with the general platform of the X-30 single stage orbital vehicles. The description fit a delta-shaped lifting body, with twin outboard vertical fins and underside rectangular engine inlets.[24]

Hypersonic aerodynamics are understood today as many test vehicles have given us good data in this regime of flight and of course space vehicles have added actual experience. Hypersonic vehicles actually ride the shock waves they create. Underside engine inlets would provide the most efficient inlet airflow for flight in this new shockwave regime. Outboard mounted vertical fins, as reported by observers, are also aerodynamically consistent with hypersonic flight.

Many of the sighted vehicles have given off a pulsating sound against a constant roar. This has given rise to belief that Aurora may be powered by pulse detonation wave engines (PDWE). Credence to this view was added when observers sighted evenly spaced puffs or "doughnuts on a rope" and the "diamonds" associated with hypersonic exhaust flow behind the suspected Aurora aircraft.[25]

PDWEs are a high-tech and practical example of the ramjet principle. A PDWE uses the created shock waves to compress the fuel-air mixture and with combustion can enormously increase the shockwave energy and therefore produce high thrust for very little engine weight. Interestingly, PDWEs can use outside atmospheric air or an onboard oxidizer giving them both an atmospheric and space capability.[26]

Hypersonic speeds are so great that aircraft skin temperatures are extreme. A structural material like the titanium ceramic composite planned for the X-30 may be used. This material offers light weight, structural strength and outstanding heat tolerance. Engineers also speculate that cryogenic fuel/oxidizer is used as a heat sink to cool skin temperatures.[27]

A fighter design using this technology would have a whopping potential. Speeds in space would be limited only by available power and range limited only by available power and life support materials. Missile and laser weapons should perform exceptionally in this environment. Given the right powerplant and weapons, ICBM interception would be possible. It boggles the mind.

THE NEXT GENERATION

Coming back to earth, let us now take a look at the next generation of aerodynamic fighters in development and coming on line operationally around the world:

THE LOCKHEED F-22 RAPTOR

The F-15 will be 20 years old this year (1995). The F-16 is a couple of years younger. They are both getting "long in the tooth". The Advanced Tactical Fighter (ATF) competition ended in a victory for the Lockheed (General Dynamics/Boeing) F-22 over the Northrop (McDonnell Douglas/Grumman) YF-23 in 1991 . The F-22 has a number of new and outstanding features:

• State-of-the-art, air-to-air missile systems (AAM), radar and IR, fire and forget.

• Very high thrust to weight ratios (better than the one to one of the F-15).

• Supercruise – the ability to cruise supersonically in military power (w/o afterburning).

• Outstanding maneuverability at all speeds -- low wing loading, relaxed stability design controlled by computers and combined with vectored thrust.

• Stealth characteristics – the most stealthy air-to-air fighter in history.

• Advanced computers controlled by fiber optic links for controls and weapons systems.

• STOL characteristics – very short takeoff and landing capabilities and ASTVOL is being tested.

• Outstanding beyond visual range (DVR) identification. Look down, shoot-down radar. IFF. Laser ranging and modern AGMs carried internally/externally.

• Outstanding visibility and a 20 MM gun.
• Self-contained ECM capability to meet current threats. [30]

Currently, the F-22 costs $35 million dollars each. (Costs are almost sure to escalate). That should limit their numbers. In my view, to extend and delay the production of this aircraft is a mistake. We most probably will not have the best fighter weapons system until the F-22 reaches operational status, and at the present pace of development, it may not be the best system when it finally reaches service. Also, if history is any judge, we should have ordered the YF-23 too, with a different engine (remember the Brewster Buffalo and Grumman Wildcat?). A modern fighter today is a budget buster, two would really be a killer, but can we afford not to buy both? Oh yeah, I forgot the cold war is over and the threat is gone. That's why India, China, and Iran are buying Russian technology as fast as they can get it, including their top fighters, which are pushing or surpassing our present fighters in performance right now.

Incidentally, we won the "Cold War" economically. Most experts predict China will have the largest economy in the world by the first decade in the new millennium which is only five to ten years away. It does not appear we can win another "Cold War" the same way. Half the world's population is still Communist controlled. China is expanding its military at the highest rate in the world right now. Why?

SUKHOI Su-35

The Sukhoi Su 35 is a refinement of

the Su-27. The Su-27 was originally designed to outperform the F-15, and was redesigned from scratch when it failed to accomplish that. The Su-35 is an improved Su-27 with *vectored thrust*.[31] As redesigned, it should outperform the F-15 quite handily and the R-73 (AA-11) IR missile is currently the best in the world (unless the Israelis have improved it!). Seventy-two Su-27s were sold to the Chinese and the Su-35 will go into squadron service in the latter half of this year with the new Russian Commonwealth Air Force. The F-22 is not expected to be operational until 2005. War in the year 2000 anyone?

THE RUSSIAN STEALTH FIGHTER -- THE MIKOYAN 1.42 MULTI-ROLE FIGHTER (MFI)

The Russian MFI fighter, believed to be designed by Mikoyan, is shrouded in secrecy, but from rumors and speculation it appears

Left: *Russia's role in aircraft design an enigma. They must sell their expertise to survive. One of the latest designs is this Sukhoi Su-35.*

design is the Vympel R-77 medium range AAM which is designed for internal carriage. This state-of-the-art missile appears to be designed for the Mik-1.42 which is consistent with Russian design practice of each new aircraft getting a new missile system. The Russians rarely retrofit their older fighters with newer missiles while western nations upgrade across the board.[33] The Russian practice has the advantage of fully integrating the new systems from scratch, but the western experience is sounder logistically and if upgrades are skillfully done, the Russians should not realize too much of an advantage.

The 1.42 looks like a complete package and reports indicate all of the features we have been discussing. The Soyuz engines will provide two-dimensional vectored thrust and the Zhuk-PH has modern synthetic aperture radar, moving-target indicator, and terrain avoidance modes. The 1.42 is evidently very expensive and Russian designers are talking about a mix with 70-80% lighter and less expensive aircraft. Configuration appears to be swept-wing and large forward canards, but this is not certain. Mikoyan has a history of mixing configurations and selecting the

to be stealthy and designed to equal or surpass the F-22. Reports indicate the Mik-1.42 is powered by two 180kN Soyuz R-79 engines and has a Phazotron Zhuk-PH phased array radar which is now used on the Su-30 and Su-35. Two engine design appears to have been dictated by the numerous operational losses of

Mig-21s with its single engine. Mig designer Rostislav Belyakov declined to discuss details, but did state "any aircraft delivered around the turn of the century that does not fully incorporate stealth does not have a chance of success".[32]

Another indication of a stealthy

best configuration based on later development and even operational experience (remember WW II where operational testing was done in combat). Russian designers are adamant in the philosophy that stealth will not compromise performance.[34] The 1.42 design appears awesome and must be followed closely. The Russia design bureaus, though broke, are very active. Sukhoi has an experimental Su-37 that features two engines, forward sweep, forward canards, and stealth. U.S. research on forward sweep has largely been discontinued.[35]

NEFMA EUROFIGHTER 2000

An agile, STOL capable fighter powered by two Turbo-Union engines of 20,250 pounds thrust in AB, the Eurofighter has a full delta wing with canards, and fly-by-wire controls. It was designed for NATO to replace the Panavia Tornado and aging F-104. Top speed is 1.9 mach and it is armed with AIM-120 AMRAAM and AIM-133 AAM missiles, as well as a 27 MM internal cannon. Models for reconnaissance, air-to-ground, air-to-air, and training are planned. Built by a consortium of British, German, Spanish, and Italian companies,[36] several prototypes are flying and were displayed at the Paris Air Show in June 1995.

DASSAULT RAFALE C

The Rafale is another delta design with canards, relaxed stability for improved maneuverability and has improved stealth characteristics. It is powered by two M-882 Snecma turbo-fans with 11,023 #s thrust dry and 16,402 #s thrust in AB. The Rafale can carry up to eight Mica AAMs and a 30 MM cannon for air intercepts and 17,637 #s of ordnance for air-to-ground operations. Top speed is Mach 2.0 at altitude and 1.2 Mach on the deck. There are one and two-seater versions as well as a naval aircraft version for carrier operations. Both the L'armee de L'air and Aeronave are purchasing these fighters in quantity, the first went into service in 1993.[37]

SAAB GRIPEN

The Gripen is also a delta design with forward canards – this one using a relaxed stability design, STOL performance, and redundant fly-by-wire computer controls. Designed to replace all three versions of the Viggen and the remaining Draaken Swedish fighters, the Gripen used the light fighter approach (weighs half the weight of a Viggen). It is powered by a single GE/Volvo Flygmotor RM-12 (404-400) of 12,250 #s dry and 18,100 #s in AB. No explicit performance data is available, but the Viggen was highly maneuverable and capable of mach 2.1 performance and the Gripen replaced it. The Gripen is the only next generation fighter that uses only one engine, but it is state of the art in every way and exceedingly versatile.[37] It should be remembered that in the late 1970s, that the F-16 was the only single-engine design among the teen fighters and its career has been very successful.

SUMMARY

Only time will tell what the future will bring for fighter aviation. The only thing certain is change. The nation that correctly identifies the proper fighter technological trends will achieve aerospace superiority /supremacy in any given timeframe. I fervently pray that achievement will be made by the "Good Guys". We have done pretty well over the last 50 years, but right now the next 10 years or so, looks a little scary.

NOTES

1. Thomas Flaherty, *Air Combat; The New Face Of War*, Time Life Books, Alexandria, Va., 1990, pp. 145-147.
2. "Washington Outlook", *Aviation Week And Space Technology*, June 12, 1995, p. 37.
3. Op. Cit.; pp. 142-143.
4. "Turning IR Missile May Be Overrated", *Aviation Week And Space Technology*, October 16, 1995, p.38.
5. "Python Capable, But Not For U.S.", *Aviation Week And Space Technology*, Oct. 16, 1995, p.49.
6. David Hughes, "Luftwaffe Mig Pilots Effective With Archer", *Aviation Week And Space Technology*, October 16, 1995, p. 39.
7. "Russians Testing Improved Archer", *Aviation Week And Space Technology*, October 16, 1995, p. 40.
8. Ibid.
9. Op. Cit.; pp. 43-44.
10. Jeffrey Ethell, and Alfred Price, *Air War South Atlantic*, MacMillan Publishing Company, New York, 1983, p. 178.
11. Op. Cit.; p. 178.
12. Op. Cit.; p. 7.
13. Op. Cit.; p. 176.
14. Green and Swanborough; Op. Cit.; p. 154.
15. Mike Spick, *BAE/McDD Harrier*, Smithmark Publishers, New York, 1991, p. 42.
16. Op. Cit; p. 607.
17. Phillip Handelman, *Beyond The Horizon Combat Aircraft Of The Next Century*, Airlife Publishing Ltd., Shrewsbury, England, pp. 67-69.
18. Article: "Lockheed ASTOVL Tests Start In June", *Aviation Week*, May 8, 1995, p. 24.
19. Handleman, Op. Cit.; pp. 93-100.
20. Op. Cit.; pp. 100-109.
21. Op. Cit.; pp. 84-85.
22. Op. Cit.; pp. 80-82.
23. Op. Cit.; pp. 86-92.
24. Ibid.
25. Ibid.
26. Ibid.
27. Ibid.
28. Ray Braybrook, *Supersonic Fighter Development*, Haynes Publications Inc., Newberry Park, CA., 1987, pp. 158-160.
29. Green & Swanborough, *The Complete Book Of Fighters*, Smithmark Publishers, New York, 1994, p. 350.
30. Handelman, Op. Cit.; pp. 22-29.
31. Green and Swanborough, Op. Cit.; pp. 556-557.
32. *International Defense Review*, June 1995, p 42.
33. Ibid.
34. Ibid.
35. "Aviation Week and Space Technology, August 3, 1998, P. 64.
36. Green & Swanborough, Op. Cit.; p. 196.
37. Op. Cit.; pp. 162-163.
38. Handelman, Op. Cit.; pp. 46-53.

CHAPTER 14

SUMMARY AND CONCLUSIONS

GENERAL:

Between 1914 and 1998, the airplane evolved into a sophisticated weapon of war and the key in that evolution has been the fighter aircraft. As a fighter pilot, I must confess to some prejudice. I love fighter airplanes. Perhaps Hemingway said it best when he wrote:

"You love a lot of things if you live around them, but there isn't any woman and there isn't any horse, not any before nor any after, that is as lovely as a great airplane. And men who love them are faithful to them even though they leave them for others. A man has only one virginity to lose in fighters, and if it is a lovely plane he loses it to, there his heart will ever be."

However, it is not my prejudice that speaks, but the historian that points out that since the first air combat, the fighter has been the most important adjunct to air war. It was the fighter in WW I that prevented enemy observation of friendly forces and allowed friendly observation of enemy. And in all the wars that followed, it was the fighter that provided the air superiority or supremacy that made victory possible. In the thirteen previous chapters, I have outlined how this came about and became history. It

---

is important to summarize those facts and to re-emphasize the vital role of the fighter in the overall scheme of war in the 20th Century.

SUMMARY:

As we have seen, it started in WW I when it became necessary to protect the observation planes providing reconnaissance and directing artillery. Everyone began arming their aircraft. It became obvious that if armed aircraft were to be effective, they had to be fast enough to catch the observation aircraft and maneuverable enough to get close enough to shoot them. After some experimentation, the machine gun became the weapon of choice. After some really "hairy" experiences with flexible guns, it was determined that the simplest way to avoid shooting yourself down was to fix the gun along the longitudinal axis of the aircraft and to aim the entire ship. The fighter was born and the "era of dogfighting" began where fighters circled and maneuvered in three dimensions to try and reach an opponent's exposed areas and his unarmed tail while avoiding the opponent's "fangs", his forward firing guns.

When tractor aircraft (the propeller in front) proved to be the most nimble type, engineers had to develop a method to safely fire a machine gun through the prop. After the Germans developed the first reliable synchronizer, they had a technological advantage for about a year. With the installation of a synchronizer, the true "fighter weapons system"

was born and the Fokker Eindecker ruled the skies until the Allies developed a superior fighter to overcome it. This pattern of a dominant fighter was repeated again and again as each side would develop a better fighter. "Air superiority" and "air supremacy" became the key to all air operations. Air superiority or supremacy went to the side who had the best fighter in sufficient numbers to dominate the air.

Very quickly the importance of air superiority to ground operations became apparent. Observation from the air provided intelligence and also could provide accurate adjustment of artillery fire which could be crucial in ground battles. Large numbers of German Eindeckers were first employed over Verdun in 1916 to gain air superiority/supremacy. They were initially successful and the French were decimated on the ground by accurate German artillery fire and observation of their every move by the Germans. Once they had supremacy in the air, the Germans had fire superiority and ascendancy on the ground. Realizing the primacy of air superiority in the campaign, the French countered with large numbers of their new and more maneuverable Nieuports, regained air superiority and ended the "Fokker Scourge". The French control of the air over the battlefield during the early summer of 1916 was crucial in finally stopping the German offensive at Verdun.

It should be noted that the Nieuport regained air superiority without a synchronizer. The Nieuports of 1916 were armed with a Lewis gun on the top wing. Many of the Allied Aces of that day flew their aircraft directly under their adversary, tilted their Lewis up and shot them down. Both the French and British flew Nieuports during this era. They solved the air-to-air gunnery problem by flying in so close they could not miss and then firing. (A technique that still works today.)

During the summer and fall of 1916, the Allies retained air superiority with the Nieuport 11 and 17 tractors, and the DH-2, and Fe-2b pushers. Neither Nieuport had a synchronizer and the pushers did not need one. The pushers were aerodynamically limited with a prop that had to be mounted amidship. In the early winter of 1916-1917, the Germans countered with the newer Albatros, Halberstadt, and Pfalz fighters. All were tractors with synchronizers and at least two guns. The air superiority tide swung back to the Germans culminating in what the British called "Bloody April" where Allied (particularly British) losses were inordinately high. The combat life of green British pilots during this period was measured in minutes in the air and in several days at the front. Much of this German dominance during this period was due to the superiority of the German fighters, but the dominance was also due to the superior fighter doctrine developed by Oswald Boelcke. Boelcke developed common-sense fighter tactics we still use today. These tactics provide for mutual support between fighters on the defense and in the attack.

When the synchronized Bristol Fighter, SE-5/5A, Sopwith Camel, and Spad appeared in large numbers in the summer of 1917 however, air superiority switched back to the Allies where it was to remain until the end of the war. The Fokker D VII, which appeared in 1918 was too late to redress this Allied superiority. As a consequence, the Allies had all the advantages of air superiority/supremacy until the armistice. It is important to remember that the air war in WW I was dictated by the fighter's dominance of the air – a pattern repeated in every war since.

However, a new element entered the picture in WW I. One of the first uses of airplanes was to drop ordnance on an enemy. With increases in speed, range, and payload, bombers appeared which were capable of striking key targets from the air when enemy fighters were not around. Attacks by surprise, at night, or when one had air superiority, were usually successful and they captured the imaginations of visionary airmen. These imaginations were further inflamed by the emotional reaction of the civilian populace to being bombed. The realities of attacks on the civilian populations were terrifying to contemplate and their morality was debated vigorously. There was much hysteria in the media about bombing attacks during WW I even though casualties were very light. Most of this hysteria was proven to be ill-founded in WW II.

By the end of WW I, visionaries had become captivated by the potential of the bomber as a new weapon of war. This fascination was to a large extent based on the difficulties in intercepting the bomber and the possibilities of the bomber overflying the traditional trenches and fortifications. The speed and celerity of the airplane and its relatively long range made the successful interception by fighters difficult and in many cases almost impossible. If a faster fighter successfully intercepted the bomber, it usually was curtains for the bomber, but interception was the rub.

A successful interceptor had to know the time, place, launch point and the direction of attack. Since the attacker moves in a three dimensional medium, the defender also had to know the altitude of the attacking force. Then the defender had to reach that altitude before the attacking force released its bombs onto the target. Since the heading, altitude, and speed of the attacker were usually unknown and subject to change at the will of the attacking commander, this was a major problem even in clear weather. In the 1920s at night or in clouds, it was impossible.

With first-hand knowledge of these factors, Douhet, Trenchard, and Mitchell virtually exploded with air power recommendations tied to bombers. Douhet's book, Command of The Air was published in 1921; Mitchell proved that bombers could sink battleships and fleets by 1923; and Trenchard became the first Air Marshall of the Royal Air Force in 1917. Although aware of WW I experience with fighters and air superiority, all three felt the bomber would be pervasive. All had real problems with their ground superiors in terms of air doctrine and how that doctrine would affect surface warfare. The debate raged around the changes that these men knew air power would dictate to the employment of surface forces in the future. Short-sighted surface officers dug in their heels as air power challenged them for scarce defense dollars. Dedicated to their arm, both Douhet and Mitchell made exaggerated claims for air power which further damaged their credibility. Both were looked on as fanatics. They also largely ignored the importance of fighters (although Mitchell had used them with great skill in 1918).

Both Douhet and Mitchell predicted Air Forces would do away with armies and navies. Douhet fervently believed that bombing would destroy the will of the people to fight. He also preached that bombers would gain "command of the air", which he saw as necessary, not with fighters, but by bombers bombing enemy airfields. Enemy air would be destroyed on the ground since there would be insufficient early warning and it would always be possible to catch them on the ground, many by surprise. The Germans built their air forces in WW II based on this Douhet corollary. This basis led to a de-emphasis of the fighter in the Luftwaffe and was to cost Hitler dearly during the Battle Of Britain and in the development of the first jet fighters.

As aircraft technology improved in the 1920s-30s, this belief in bombers was exacerbated by technology. During the late 1920s and early 1930s, as all metal monoplanes appeared, the bombers outperformed the fighters of that day. With increased altitude capabilities, the time from early-warning to interception was reduced substantially and as many of the bombers were faster than the fighters, the fighters had to reach altitude before the bombers arrived to be successful. This led

Stanley Baldwin, the Prime Minister of Britain in 1932 to state "the bomber will always get through" and consequently the everyday citizen would face obliteration from the air.

The Spanish Civil War spanned this era. In its early days, fast bombers were able to escape from the biplane fighters of the day, but with the introduction of modern monoplane fighters such as the I-16, and Me-109, those days were over. In Spain and China, unescorted bombers were meat on the table if intercepted by modern fighters – unless air superiority/supremacy had been established previously. The Germans and Italians got to test their air power ideas in the Spanish Civil War. The efficacy of fighters immediately stood out as air superiority vacillated back and forth depending on who had the dominating fighter. First the Republicans dominated with the Russian I-16, then the pendulum swung back to the Nationalists as the Me-109 appeared.

In Spain, the Germans solidified the close air support so essential to their "Blitzkrieg" ground attacks in WW II and refined revolutionary fighter tactics first developed by Oswald Boelcke in WW I. The Rotte (pair) and Schwarm (supporting second pair) resurrected by the German Condor Legion in Spain provided for pairs to fly line abreast in loose formation that allowed the four aircraft to mutually support each other. This formation was far superior to the three ship Vics in vogue at the time, and was adopted by almost all of the air forces of the world during WW II. It is still the basis for current combat fighter formations.

Fighter armament began to change

during this period. Combat brought out the advantages of more and heavier guns. The need for more guns to quickly knock down opponents lead to remotely mounted guns (i.e., the Spitfire and Hurricane had eight .303 caliber machine guns; the volume of fire which was far too complex to synchronize). Fortunately, electrical solenoids provided a way to fire guns remotely from the cockpit. This led to guns being installed in the wings outside the arc of the propeller and their being installed to converge on a target at a specific range that was effective for the particular gun. With no synchronizers required, the guns were programmed to fire at their maximum design rates. Larger calibers and higher rates of fire were employed by many new designs. For example, in addition to the eight gun Hurricanes and Spitfires, the Me-109E had two machine guns synchronized to fire through the prop and two 20 MM cannons in the wings. Late models of the I-16s in Spain were equipped with two high velocity, rapid-firing 20 MM cannon in their wings and two rapid-fire machine guns to fire through the prop. These higher rate of fire guns and cannon gave these I-16s more weight of fire than either the Spitfire or Me-109.

As demonstrated by the Spanish Civil War, the fighters had caught back up and surpassed the bombers in performance by the mid-to-late 1930s. But early-warning remained a problem and was particularly vexing to fighter adherents. The U.S. Army Air Corps experimented with ground observer corps to locate incoming air attacks. Claire Chennault, as a proponent of fighter aviation, was particularly taken with this idea. After Chennault resigned from the US Army, the USAAC tested the idea with good results in 1938 maneuvers. By this time, Chennault was practicing what he had preached in China. His "Coolie network" provided early warning of Japanese raids very successfully and even his obsolescent old Curtiss Hawk fighters were very successful against unescorted Japanese bombers when they intercepted them, as they often did. Limited to sound and visual contact, ground observers obviously were less effective in bad weather and darkness. Conversely, Japanese bombers rarely caught Chennaults fighters on the ground which speaks well for his "eyeball" early warning net.

The tie-breaker for fighters however, appeared in Great Britain in the late 1930s with the invention of aircraft radio detection and ranging (RADAR). Tied together with good radio communications and viable Operations Centers, this electronic system made it difficult for an aggressor to attack a defender from the air without warning. Radar had the advantage of being able to detect incoming aircraft in clouds or darkness where visual contact was impossible.

Radar provided early-warning, direction, altitude, and made it possible to mass defensive fighters to meet the attack. It was a capability of which Douhet and Mitchell never dreamed and made it extremely difficult for the Germans to catch the British fighters on the ground. In previous campaigns in Poland, Norway, and France, the Luftwaffe had gained air superiority by bombing fighter airfields and catching the fighters on the ground. In Britain, it was a whole new ballgame.

Over Britain, every raid was contested as the British air defense system met them with swarms of Spitfires and Hurricanes. German intelligence never identified key fighter operations centers and assumed these centers were fortified underground (therefore very poor targets). Even so, if they had persisted in their attacks against British fighter fields, they very well might have prevailed as sector operations were controlled from above-ground facilities at several of these fields. The British ability to mass their fighters at decisive points was crucial to their success in the Battle of Britain.

The Germans discovered very quickly that the only aircraft capable of surviving by itself over Britain was the single-engined Me-109. Without Me-109 escort, none of the other Luftwaffe aircraft could survive. However, the Me-109 was the only operational single-engine fighter the Germans had in 1940 and its combat radius was so small it could not meet Luftwaffe requirements. With its limited range, it could not operate north of London and had only 15 minutes endurance south of London. As pointed out above, the Spitfire and Hurricane were equally limited in fuel capacity. Both the Germans and British believed that fighters were restricted in performance by weight (fuel in this case).

The Japanese ran into range problems in 1937-38 in China. Their solution was more internal fuel and external, droppable tanks. They designed exceptionally light fighters to compensate for the weight of extra fuel. Their fighters were very clean, light, and had exceptional range. The first forces surprised by this capability were Chennault's Chinese fighter force. When Chennault first encountered the Japanese Claude and Nate, he withdrew his fighters into internal China to be out of range of these superior fighters. When the Japanese attacked internal Chinese targets with unescorted bombers, he savaged them with his old Curtiss Hawk fighters. The external drop tanks for the Claudes and Nates (and later Zeroes) were the Japanese answer so they could escort and protect their bombers. This innovation surprised Chennault's obsolete fighter forces and they were decimated by these now long- range Japanese fighters. The Japanese fighters' presence were crucial to bomber survival and the re-gaining of air superiority over China. Chennault knew what to do to redress this superiority, but he needed better trained pilots and better fighters. He would get both in 1941 when the American Volunteer Group, "the Flying Tigers" became operational.

Had the innovation and fighter design philosophy of the Japanese prevailed in Europe, the history of WW II would probably have been very different. For example, had the British been equipped with the Mitsubishi Zero fighter in 1940, they could have supported Allied forces in Norway with fighters from Britain. The Zero had that kind of range. Likewise, the Zero could have ranged over Britain during the Battle Of Britain had it worn the Swastika rather than the "Red Meatball". The limited range of the

British and German fighters was a critical limitation all through the war.

At the beginning of WW II, the British tried unescorted raids with bombers (the Spitfire and Hurricane had insufficient range to escort bombers into Germany). The results were disastrous for the British bombers. German fighters had a field day. After some heavy losses, the British switched to night bombing attacks only. Again, fighter range was proven to be critical.

When the U.S. entered the war, they believed that their four engined, high altitude (20,000-25,000 feet) bombers, armed with .50 caliber machine guns and massed in "combat box formations", could fight their way through unescorted to the target in daylight. The British, based on their experience, were highly doubtful and said so. The Americans, however, were trained for daylight precision bombing and to retrain for night operations was not a viable option. No fighter had the range to escort bombers into targets in Germany. A long-range fighter that could fight on equal terms with Luftwaffe fighters was desperately needed. The USAAF had no option but to "gut it out" with unescorted bombers in 1943, or give up the idea of bombing Germany. However, unlike the British and Germans who thought it impossible, they began to try to develop long-range fighters.

The first raids proved a new experience for the Germans. The Bombers' heavy machine guns proved almost as effective as the Americans had hoped and expected them to be. Initially, German fighter losses were heavy and American bomber losses were acceptable. Unfortunately, the Luftwaffe fighter pilots proved that the British knew what they were talking about as the German fighters quickly adjusted to the American heavy bombers and their tactics. When the American bombers tried to penetrate deep into Germany, their losses increased markedly and became unacceptable. The existing escort fighters, the American P-47s and P-38s, did not have sufficient internal fuel to escort the bombers to deep German targets. When jumped by German fighters, the American fighters had to jettison their external fuel tanks to survive and they had insufficient internal fuel to escort the bombers all the way to the targets and back. Unescorted American heavy bombers were soon unable to survive in air dominated by German fighters. The Germans created heavy attack squadrons of "bomber shooters" using rockets, and heavy cannon and even dropped bombs on the large American formations. These aircraft could fire on the Americans outside the range of their bomber's defensive machine guns. Losses increased even more.

In December 1943, the American P-51B Mustang was introduced in Europe. The P-51 was superior to contemporary German fighters and had the range to penetrate to German targets. The superior aerodynamics of its wing and design and the perseverance of American leadership in cramming additional internal fuel into the Mustang were crucial to the air war over Germany. The P-51s' introduction was the turning point in the air war.

When unleashed on the heavily laden "bomber shooters", the Germans were massacred by the P-51s. When Goering heard there were Mustangs over Berlin, he was quoted as saying "We have lost the War". He was right.

The heavy American daylight bombers forced the Luftwaffe defense fighters into the skies to defend Germany and the American fighter escorts shot them down. Air superiority followed quickly as the German losses mounted and their pilot skill level and experience fell dramatically. In the spring of 1944, the fighters were released from pure escort and were charged with hunting down the German fighters. This included the strafing of airfields and ground targets. Air supremacy for the Allies was achieved by May 1944. D-Day in June saw no serious German air resistance. Long-range fighters were decisive.

Fortunately, Hitler and the top Luftwaffe leaders did not understand the efficacy of fighters. If they had, and had expedited the production of the Me-262 jet fighter to a top German priority, we all might be goose-stepping today. About 100MPH faster than the P-51, the Me-262 was a quantum-level jump into the future and could have turned the war around if it had arrived on the scene in numbers six months earlier. Hitler tried to turn the aircraft into a bomber which delayed its operational delivery three or four months, but the real delay was in the fabrication of its jet engines which was very complex for its day – thank God. An all-out effort to get this aircraft operational a year earlier might have been possible and very well might have brought air superiority back to the Luftwaffe to stay.

The German loss of air superiority was largely due to the heavy loss of trained, combat fighter pilots. It is apparent that the pilots that fly the fighters are an essential element of any fighter weapons systems. During the critical phases of the Battle of Britain, Great Britain ran into this same shortage of pilots as did the Japanese after her heavy carrier pilot losses at Midway. It is apparent that combat-trained pilots are one of the longest lead time requirements in air war, particularly fighter pilots.

In the Pacific, the fighter was equally decisive. During the first six months of WW II, where the Mitsubishi Zero dominated the skies, Japan's dominance went unchallenged (except for Chennault's AVG). Once again the superior fighter, available in sufficient numbers, dominated the skies and the campaigns fought below those skies. When large numbers of U.S. P-38s, P-47s, F6Fs, F4Us, and P-51s appeared, the air war in the Pacific was essentially over.

Perhaps the best example of the efficacy of the fighter in the Pacific was MacArthur's campaign in the Southwest Pacific where he first gained air superiority using land-based fighters and fighter escorted bombers and then used amphibious operations under that umbrella to cut-off and isolate Japanese bases. He had the lowest casualty rate in WW II with this innovative method and it was totally based on the facility of his P-38 fighters to gain air superiority. Again, in WW II, as in WW I, the fighter was the key to air war.

The dominance of the fighter was re-emphasized in the conventional operations in WW II, but the dropping of the atomic bomb on Hiroshima at the end of the war reignited the fascination with bombers. If one bomber armed with nuclear weapons got through, the results were sure to be catastrophic. In 1945, it required a large bomber to carry the nuclear bombs of that day and also to provide the intercontinental range needed to deliver them to world targets. It appeared that the atomic bomber was supreme and the predictions of Douhet were finally being borne out.

But in June of 1950, the Korean war broke out and the U.S. with a near monopoly on nuclear weapons refused to use them against either North Korea or China. In the conventional war that followed the fighter was still king. But the aerodynamics had changed. We were in the jet age and the swept wings (which delayed the effects of compressibility) of the day ushered in the trans-sonic age where air became compressible and all of our previous aerodynamics (which assumed air was incompressible) had to be recalculated and tested in the air. But the tactics of Oswald Boelcke still worked. The best fighter weapons system still provided air superiority.

In the mid-1950s developments drastically affected fighters. The first was mid-air refueling, where suddenly the primary aircraft could be refueled by tanker aircraft and the fighter had almost unlimited range with no increase in size or fuel capacity.

The second was the necessity to intercept nuclear bombers which focused fighter development on the interception of high-speed, high altitude nuclear-armed bombers at the expense of a fighter-versus-fighter focus. Another development that affected fighter design was the miniaturization of nuclear weapons. This allowed fighters to carry atomic bombs and by the late 1950s, fighters were equipped to carry fusion (hydrogen) bombs.

The nuclear bomber-shooting requirement was responsible for the specialization of fighters as interceptors and their close control by ground radar (GCI). As nuclear-armed jet bombers were developed, interception timing became critical and the interceptors were assured of only one pass against the fast high-flying jets. This led to the arming of the interceptors with large numbers of unguided rockets and then guided missiles. This was an attempt to develop projectiles that with a single hit would knock-down the fast-moving nuclear bombers. The interceptors were directed by ground radar until they could take over the intercept on their own radar sets. Radar interception in all kinds of weather and at night, exacerbated the single-hit requirement as there would probably not be another chance for the attacking fighter.

Unguided rockets were salvoed

from the interceptor in large numbers. The shotgun effect of these salvoes was expected to provide at least one killing hit. As guided missiles became available they were substituted for the salvo of unguided rockets. Rockets and missiles were fired automatically by the interceptor fire control system or semi-automatically by the pilot on signal from the radar fire control system. In clouds or at night, the interceptor pilot might never see his target. Slowly but surely the guided missile replaced the unguided rocket armaments.

Heavy bombers were not expected to maneuver extensively and consequently guided missiles were not designed to be launched from a maneuvering fighter. This was to adversely affect missile effectiveness against fighter targets in clear weather. Missiles were radar guided for weather and infrared (IR) guided for clear weather. Tests indicated a very high probability of kill (Pk) for these newer air-to-air missiles so long as they were launched within missile parameters.

Since missiles were often fired at ranges that could exceed the pilot's ability to see the target in clear daylight and it was impossible to see targets in weather and at night, the need for a close-in gun attack was seriously questioned. With high missile Pks, designers began leaving out the traditional gun and sacrificing the gun/ammo payload to carry more missiles, rockets, or fuel.

U.S. air defense SAMs and interceptors were equipped with nuclear warheads to deal with large formations of bombers. Unguided air to air nuclear missiles were carried by certain interceptors. It still remains a question as to how much harm would have been done to friendlies if these weapons had been used over friendly territory. If nuclear war threatens again, we may yet get an answer to that question. The nuclear air defense capability would have to be re-created as it no longer exists today.

This missile/interceptor approach dominated the fighter world to the extent that many air defense personnel never considered that the nuclear bombers might be escorted by long-range, air-refueled fighters or that the bombers themselves might be fighters carrying smaller nukes. Either way the interceptors would have had to fight maneuvering fighters and their missiles and rockets would have been much less effective.

Nor did this interceptor syndrome have any application to tactical fighters in conventional war where the game was still fighter against fighter, but unbelievably, it got applied to them too. In the USAF, tactical fighter pilots began to scream bloody murder when the Navy F-4 interceptor was bought in the late 1950s and early 1960s without an internal gun. Not only were the F-4's missiles (AIM-7s and AIM-9s) not very effective against other fighters in maneuvering fights, they were worthless in close-in (<2,000 feet) fights. When F-4s were used as fighter-bombers, the gun was sorely missed. For ground attack, the gun was the most effective weapon against soft-skinned ground targets and remember more than 50% of the casualties caused by aircraft in WW II were caused by strafing. In 1960, it was ludicrous to remove the gun from any fighter, but both the U.S. and USSR did it.

During this time however the USAF also lost its fighter emphasis when it imitated the Strategic Air Command with the F-100, F-101, F-104, and F-105. They were all designed to carry nuclear weapons and to go fast and penetrate to targets. They were not designed to maneuver and fight other fighters. Their performance was not really suitable for fighting other fighters, as they were designed for nuclear strike. Nor were aircrews properly trained for air-to-air fights. Many long and precious fighter training hours were used practicing nuclear deliveries and nuclear escape maneuvers.

All of these "Century Series" fighters were supersonic in level flight. To achieve supersonic flight each needed to use after-burner in level flight, or to dive to accelerate past the sharp drag rise at mach 1.0. The afterburners raised fuel consumption by a factor of four or five which meant ABs had to be used judiciously. Even though they had low supersonic drag, the Century Series birds were very inefficient subsonically. When supersonic, even slight maneuvering slowed them to subsonic speeds. The drag of external stores reduced these aircraft to subsonic top speeds even though many had top speeds of better than mach 2 in clean configuration when operated at high altitudes above 30,000 feet. In combat however, SAMs forced these aircraft down to low altitudes where heavy air densities and high ambient temperatures prevented supersonic flight even in a clean configuration.

Fighting transonically against subsonic aircraft with high thrust/weight ratios, supersonic fighters were actually at a disadvantage because of their heavier weights and their high drag when maneuvering. Lighter transonic aircraft (Folland Gnats, Orenda powered F-86 Sabres, F-86Hs, Mig-17s, etc.) although without adequate thrust to exceed mach 1 in level flight(their drag at mach 1 was too high and they had no afterburners), had surplus thrust for maneuvering flight in the .85 to .95 mach range where all of these engagements occurred. The transonic birds could maneuver in this mach range without losing energy while the heavier supersonic planes lost energy when they maneuvered.

When these supersonic aircraft used their afterburners to offset this disadvantage, they found their "burners" ineffective in increasing energy immediately unless they "unloaded"; i.e., went into unaccelerated flight (this was often not practical in a fight). Unloading was effective for two reasons: it reduced drag and increased airflow through their engines which resulted in markedly increased thrust. At, or near the speed of sound, the supersonic aircraft had a lower drag than the transonic designs, but below .93-.95 mach, they lost this drag advantage.

The supersonic fighters used as fighter bombers had an added disadvantage as most fighter bombloads were carried externally, which exacerbated the drag disadvantages outlined above. In Vietnam, where SAMs forced us down to low altitudes, our fighters were also loaded down with external stores. So in a situation where the supersonic fighters were at a disadvantage if they slowed down, we forced them to slow down by the way we used them. Hence, we had the phenomenon of .95 mach Mig-17s shooting down mach 2 F-105s in Vietnam. In the Indo-Pakistani conflicts, we found .93 mach Indian Folland Gnats threatening or shooting down mach 2.3 Pakistani F-104s. This development was not foreseen by the experts.

However, the supersonic fighters could be effective when used properly. From the first days of fighter conflict, faster fighters have been able to defeat more maneuverable, but slower fighters, by maneuvering in the vertical, i.e., diving and zooming and using superior speed capabilities to dive to the attack and then to zoom safely away using the excess energy available from the higher speeds. The heavier, faster century series aircraft (including the McDonnell Douglas F-4) could accelerate to supersonic speed by unloading using the 'burner and attacking supersonically, then climbing back out of reach using this higher energy. The drawback was that vertical attacks and defenses required a high level of pilot skill and training and the supersonic birds had to be in a clean configuration. You cannot use the vertical carrying a load of bombs and supersonic designs need to use the AB efficiently. In Vietnam the F-105s and F-4s were too far from home (or a tanker) to use the 'burner extensively.

"Green" fighter pilots have the tendency to think in the horizontal. To successfully maneuver in the vertical requires that the attacking pilot must visualize the battle in three dimensions and must know how to maneuver his/her aircraft advantageously against another skilled fighter pilot and his/her aircraft. It takes time and experience to train embryo fighter pilots this skill. When you add the complexity of fighting a heavy supersonic design against a light transonic, the equation becomes even more complex.

During the Vietnam War, these facts were largely ignored. Our fighter pilots were highly experienced and had many hours of flying time, but very little time had been spent on air-to-air combat. Only after our losses to Mig-17s, Mig-19s, and Mig-21s did we begin to train properly. In addition, as we had inadequate radar coverage over North Vietnam and parts of Laos, we could not "identify friend from foe"(IFF), which deprived us of the use of our superior beyond visual range radar missiles for fear of downing a friend. The rules of engagement (ROEs) throughout the war demanded visual identification (VID) before we could fire. The lack of coordination between USAF, USN, and USMC aircraft was largely responsible for this VID requirement. Repeated attempts to establish a single airspace manager to solve this problem failed.

The VID requirement almost dictated that close-in fights would occur where a gun was needed desperately. These ROEs were blatantly stupid. Only once during the entire war was adequate separation between friend and foe achieved. Operation

Bolo resulted in seven enemy losses with no friendly losses.

In the early 1980s, however, the missiles caught up and with adequate IFF equipment became highly effective. It began with AIM-9L advanced "Sidewinder" missiles which were all-aspect capable, IR guided systems. The "L" could be fired from any aspect (angle) and its discerning, cooled, IR seeker would lock on to the target regardless if fired from the front, back, side, etc. In the Falkland Islands, the brand-new AIM-9Ls proved highly effective and were decisive in gaining air superiority for the British in that war. In the Bekaa Valley in 1983, the "L" was equally effective as were improved radar-guided AIM-7 Sparrows for the Israelis. The F-15, with its superior air-to-air radar and IFF systems, was deadly in combination with Israeli command and control aircraft. No VID was required in either war.

In the Falklands, a slower, more maneuverable Harrier, was able to outfight faster more numerous fighters as the AIM-9Ls were so superior to the older, more limited AIM-9Bs with which the Argentines were armed. Only one air-to-air engagement was attempted by the Argentines. In this single battle, Argentine Mirages attempted to maneuver in the horizontal with the Harriers instead of using dive and zoom tactics as they should have. The results were predictable. The Harriers, with their V/STOL characteristics, made admirable air-to-air missile launch platforms and their dominance, thanks to the superiority of the AIM-9Ls, was a shocker to the so-called military experts

around the world (although USN exercises in 1975 indicated the Harrier was very effective air-to-air against F-14s).

As a result of our lack of fighter maneuverability during the Vietnam years, the U.S. built a new series of fast and maneuverable fighters. The F-14, F-15, F-16, and FA-18 were all built to out-accelerate and outmaneuver the opposition. Thrust/drag and thrust/weight ratios were higher for these "teeners" than any fighters in U.S. history. All were capable of acceleration going straight up with normal payloads. The F-16 ushered in "relaxed stability" controlled by computers and "fly-by-wire" systems to gain unprecedented maneuverability. In numerous incidents, such as the Bekaa Valley and Desert Storm, these new American fighters dominated wherever they fought. The Israelis, using F-15s and F-16s in the Bekaa shot down scores of Arab aircraft with no Israeli losses. In the Persian Gulf, these successes were repeated. The dominance of the teeners guaranteed air superiority in both wars.

In Panama and the Persian Gulf, stealth fighters were used for the first times in actual combat. In modern ECM dominated air wars, these fighters proved invisible and uninterceptable. On scores of sorties, these aircraft were unscathed even against some of the most highly defended targets in the history of air warfare. The results of stealth were to take fighter-bombers back to the pre-radar eras of air warfare – a great advantage for the bomber.

In the previous chapter, we looked at the future fighters on foreign

drawing boards around the world. Those fighters are all designed to exceed the performance of the American teeners that have dominated the last 15 years. In addition, a whole new generation of air-to-air missiles is becoming operational. The American F-22 and our new AAMs may be the most important fighter weapons system designs in U.S. history.

CONCLUSIONS:

After reviewing air war history, it is apparent that the fighter has been a dominant factor in every air war since the invention of the airplane. Control of the air by one belligerent has usually guaranteed that side's victory and the key to the control of the air has been the fighter weapons system.

However, the fighter weapons system has been evolving and changing through the years. In WW I, when the machine guns were first aligned with the longitudinal axis of aircraft, the fighter was defined. The fighter was an aircraft weapons system that had to be maneuvered and the entire aircraft aimed to attack an adversary's vulnerable area, usually his tail. Eighty years later it remains substantially unchanged; however, that definition began to change when the first air-to-air missiles appeared in the 1950s-1960s. The definition underwent continued metamorphosis in the 1980s as IR guided missiles became all-aspect and radar guided missiles were improved. This continued into the 1990s with fire and forget designs in recent years.

It is clear that AAMs have the

capability to eventually redefine the fighter. The Falklands, the Bekaa Valley, and Desert Storm all demonstrated the increasing ascendancy of AAMs. If those conflicts are any example, the fighter equipped with the best missiles will prevail in the future. It is the best weapons system that will win.

Analyzing current technology and recent history, fighters alone may have already seen their heyday. To put it succinctly, if given a choice today, between the best AAM or the best fighter, knowledgeable airmen would probably choose the missile. Yet while we are spending billions on new fighters, our research and development for new and improved missiles is underfunded. For example, many believe the Russian AA-11 is currently the best operational IR missile in the world and U.S. leaders are dragging their feet on purchasing the AIM-9X which is said to be better than the AA-11. The British ASRAAM and Israeli Python also must be comsidered.

One should also evaluate missile design philosophy. Is our philosophy in building one missile system for all of our fighters across the board better than the Russian practice of integrating the missile system with new aircraft? Does one dictate compromises versus the other? The best missiles will win. There is no second place.

In addition, ECM and Stealth features that might protect our fighters against missiles are struggling to get funding as well. Our emphasis must be on the total fighter weapons system. However, judging future requirements from the recent past, if one had to choose between capabilities, the smart person would choose outstanding missile performance over outstanding fighter performance (consider the Harrier versus the Mirage in the Falklands). Conversely, defense against missiles – stealth and ECM should be chosen over outstanding fighter performance for night and all-weather fighters (no F-117 was hit in Iraq during Desert Storm).

Protection against SAMs must have priority as well. The shoot down of Scott O'Grady's F-16 is case in point. For whatever reason, his aircraft had inadequate ECM protection from SA-6 SAM missiles.

To paraphase General Carl "Tooey" Spaatz: A second best fighter weapons system, like a second best poker hand, ain't worth a damn. It is my proposition that a second best fighter weapons system could cost us the next war. If we are to prevail, our fighter weapons systems must be second to none.

Above: *The lessons of history are clear. If the U.S. is to maintain its superiority/supremacy in the new millennium we must buy the F-22 Raptor as soon as possible.*

AIR REFUELING

APPENDIX ONE

No advance in aircraft performance has had more effect than the development of Air Refueling. No longer do aircraft have to be inordinately large to possess long range. As long as the planes are operational and tanker aircraft are available– only aircrew endurance limits aircraft range.

This has been particularly true for fighters. Since the mid 1950's, USAF & USN fighters have been deploying world wide without landing.

One of the keys to modern airpower has become the availability of tanker support. To date these tanker aircraft have not been targeted by enemy forces. In future wars, I believe they will be.

Airpower strategists must carefully reconsider this air power revolution. Air refueling has been key to our successes in Vietnam, the Gulf War, and contingencies everywhere. We must expect our tanker support to be attacked in the future, and we must plan to attack the tankers used by our enemies. The arithmetic is simple – if one of our tankers supports eight fighters, and the tanker is lost, it will cost us both the tanker and the eight fighters as well as any future use of the tanker. The same is true for any enemies. It's more bang for the buck.

The next page illustrates some of the history of fighter refueling in the USAF.

AIR REFUELING
A REVOLUTION FOR FIGHTERS
APPENDIX ONE

1. Practical air refueling began in the mid-1950s in England using the probe and drogue method shown below. Shown is an F-100F (two seater) refueling in 1959 in the Philippines with a KB-50J tanker. Drogue refuelers usually can refuel three fighters at a time (one on each wing–one on the tail drogue). Note the straight probe of this F-100F. Contrast it with the later curved probe on the F-100Ds pictured on page 110.

2. Boom refueling was adopted as standard by the U.S. Air Force in the 1960s. Developed by the Strategic Air Command (SAC) for SACs bombers, it was easier to transfer the required larger amounts of fuel to the bombers using the boom method.

3. Boom refuelers could be adapted to probe/drogue refuelers–as shown here. Note the A-7 refueling here is a U.S. Navy model. The U.S. Navy and Foreign nations still use the probe and drogue refueling method.

THE ROLE OF FIGHTER AIRCRAFT IN WAR

APPENDIX TWO

DEFINITIONS

AEROSPACE POWER

Military power provided by the use of aerospace vehicles and their associated weapons systems within the earth's atmosphere and surrounding space. To include all atmospheric and space vehicles, i.e.; airplanes, helicopters, balloons, dirigibles, blimps, manned and unmanned rockets, satellites, missiles, etc.

AIR POWER

The military power provided by the use of air vehicles within the earth's atmosphere, i. e.; airplanes, helicopters, balloons, dirigibles, blimps, missiles, etc.

AIR SUPERIORITY

The control or the ability to gain temporary control of the airspace over a certain area for a period of time.

AIR SUPREMACY

The dominant control of airspace over a given surface area. Implies the control to use the airspace for any desired mission.

CLOSE SUPPORT

Direct air support for surfaces forces in contact with enemy ground forces

INTERDICTION

The attack of enemy ground force lines of communication and supply. Ideally the capability to isolate the enemy on the battlefield.

STRATEGIC AIR

The direct attack on a nations means of making war, i.e.; industries, utilities, government transportation, work force, etc.

A SELECTED BIBLIOGRAPHY

Anderson, Clarence B.; *To Fly And Fight*; R.R. Donnelly & Sons; Crawfordsville, IN.; 1990.

Allen, Peter; *The Yom kippur War*; Charles Scribner's Sons; New York; 1982.

Batchelor, John and Cooper, Brian; *Fighter, A History of Fighter Aircraft*; Scribner and Sons; New York; 1973.

Belotte, James and Belotte, William; *Titans Of The Seas*; Harpers Row, Publishers; New York; 1975.

Bishop, William A. (Billy); *Winged Warfare*; Arco Publishing,Inc.; New York; 1967

Burrows, William E.; *Richthofen A True History Of The Red Baron*; Harcourt, Brace & World; New York; 1969.

Caldwell, Donald L.; *JG 26 Top Guns of the Luftwaffe*; Ivy Books; New York; 1991.

Clancy, Tom; *Fighter Wing*;Berkely Books; New York; 1995; ISBN Unk.

Closterman, Pierre; *The Big Show*; Ballantine Books; Random House; New York; 1951.

Cohen, Eliezer "Cheetah"; *Israel's Best Defense, The First Full Story Of The Israeli Air Force*; Orion Books; New York; 1993

Deighton, Len; *Fighter The True Story Of the Battle Of Britain*; Jonathan Cape; London; 1977.

Dunn, William R.; *Fighter Pilot: First American Ace of WW II*; The University Press Of Kentucky; Lexington, KY.; 1982.

Epic of Flight, three volumes: "The Aeronauts", "The Road to Kitty Hawk", and "The First Aviators"; Time-Life Books; Chicago; Il.; 1981.

Ethell, Jeffrey and Alfred Price; *Air War South Atlantic*; MacMillan Publishing Company; New York; 1983.

Fonck, Rene; translated by Martin Sabin and Stanley Ulanoff; *Ace OF Aces*; Ace Books; New York; 1967.

Forrester, Larry; *Fly For Your Life The Story Of R.R. Stanford Tuck*; Bantam Books; New York; 1981.

Foss, Joe and Brennan, Mathew; *Top Guns*; Pocket Star Books; New York; 1991.

Freeman, Roger A.; *Zemke's Wolf Pack*; Orion Books; New York; 1988.

Galland, Adolph; *The First And The Last*; Balentine Books; New York; 1954.

Goodson, James A.; *Tumult In The Clouds*; St. Martin's Paperbacks; New York; 1983.

Griess, Thomas E.; *The Second World War, Europe and the Mediterranean*; The West Point Military History Series; Avery Publishing Group; Wayne, NJ.; 1989.

Gunston, Bill & Peacock, Lindsay; *Fighter Missions, Modern Air Combat-The View From The Cockpit*; Crown Publishers Ltd.; New York; 1989.

Hallion, Richard P.; *Storm Over Iraq Air Power And The Gulf War*; Smithsonian Institution Press; Washington, D.C.; 1992.

Hallion, Richard P.; *The Naval Air War In Korea*; Keningston Publishing Co.; New York; 1986.

Hallion, Richard P.; *Rise Of The Fighter Aircraft 1914-1918*; Nautical & Aviation Publishing; Baltimore, Md.; 1984.

Hallion, Richard P.; *Strike From The Sky*; Smithsonian Press; Washington, D.C.; 1989. ISBN # 0-87474-452-0.

Hastings, Max and Jenkins, Simon; *The Battle For The Falklands*; W. W. Norton & Co.; New York; 1983.

Herzog, Chaim; *The Arab-Israeli Wars*; Random House Inc.; New York; 1982.

Higham, Robin and Siddall, Abigail; *Flying Combat Aircraft of the USAAF-USAF*; Iowa State University Press; Ames, Iowa; 1975.

Higham, Robin and Williams, Carol; *Flying Combat Aircraft of the USAAF-USAF*, Vol 2; Iowa State Univesity Press; Ames, Iowa; 1978.

Isby, David C.; *War In A Distant Country Afganistan: Invasion and Resistance*; Sterling Publishing Co.; New York; 1989.

Jablonski, Edward; *Man With Wings*; Doubleday & Company; Garden City; New York; 1980.

Johnston, J.E.; *Full Circle*; Chatto And Windus; London; 1964.

Johnston, J. E.; *The Story Of Air Fighting*; Bantam Books; New York; 1985.

Jones, Lloyd S.; *U.S. Fighters*; Aero Publishers; Fallbrook, CA.; 1975.

Kennett, Lee; *The First Air War*; The Free Press; New York; 1991.

Lundstrom, John B.; *The First Team*; United States Naval Institute; Annapolis, MD.; 1984.

Mckee, Alexander; *Strike From The Sky*; Lancer Books; New York; 1960.

McCudden, James T.B.; *Flying Fury*; Ace Books; New York; 1968.

Miller, Thomas G.; *The Cactus Air Force*; Bantam Books; New York; 1969.

Nordeen, Lon O. Jr.; *Air Warfare In The Missile Age*; Smithsonian Institution, Washington, D.C.; 1985.

Overy, R. J.; *The Air War 1939-1945*; Stein and Day; New York: 1980.

Piekalkiewicz, Janus; *The Air War 1939-1945*; Sterling Publishing, New York; 1978.

Rickenbacker, Edward V.; *Rickenbacker*; Fawcett Crest; Greenwich, CN.; 1967.

Ries, Karl; *The Legion Condor*; Ballantine Books; New York; 1992.

Sims, Edward H.; *Fighter Tactics and Strategy 1914-1970*; Harpers Row; Publishers Inc.; New York; N.Y.; 1972.

Stahl, P. W.; *KG 200 The True Story*; Jane's Publishing Company Ltd; London; 1981.

Tolliver, Raymond and Constable, Trevor; *Fighter General The Life Of Adolph Galland*; AmPress; Zephyr Cove, NV.; 1990.

Townsend, Peter; *Duel Of Eagles*; Simon and Schuster; New York; 1972.

Udet, Ernst; *Ace Of The Iron Cross*; Arco Publishing, Inc; New York; 1981.

Wyden, Peter; *The Passionate War*; Simom & Schuster; New York; 1983.

EPILOGUE

Since I completed the text of *Wings That Stay On*, a concerted effort to challenge the F-22 program has arisen in the U.S. Congress. The F-22 has had the highest priority of any program in the USAF for several years and, for the reasons outlined in the book, this priority is well-justified. The challenge from Congress was unexpected, but conventional, and understandable. State-of-the art technology is always expensive, and the F-22 is a quantum-leap forward in capability, and therefore costly. In addition, the demise of the once powerful Soviet Union and the Cold War has decreased the motivation to maintain the air superiority the U.S. has had over the last 50 years.

In addition, the Clinton administration has challenged the USAF with peace-keeping missions all over the world, while markedly decreasing forces in-being. This has put a heavy strain on the USAF, as well as the other armed forces, to do more with less. Airlift, Electronic Countermeasures (ECM), Reconnaissance, and Strategic Forces all have had shortfalls, and are in stiff competition for the defense dollar. The loss of the all-important USAF ECM aircraft; the EF-111 and F-4Gs, with no dedicated follow-on, is a case in point. Navy EA-6s have not been able to meet the ECM requirements for both the USAF and USN. This has caused some vulnerabilities to missiles in Bosnia and Iraq. Also, the toll on both aircrews and aircraft from constant world-wide deployments has been heavy. The retention of highly trained aircrews and maintenance technicians is at an all-time low. The costly F-22 program, and the follow-on Joint Strike Fighter (JSF) thus have endured heavy scrutiny. The F-22 program has been continued by a hair's breath, and may still be threatened in the future.

It remains my belief that both the F-22 and the JSF are *vital* to the security of the United States. Our capability to control the air is as important as ever, and the whole world is nearing parity with our F-16 and F-15, on which we have depended for the last 20 years. A small country like Greece has just opted to buy the Eurofighter, and Japan the F-2, both equal, or superior to our current operational fighters. Russia, China, and India are re-equipping with the latest Mig and Sukhoi designs. We must maintain our technological superiority, or lose a war in the future. Losing a war is far more costly than any option available to us now.

In contrast to many members of Congress, I believe the threat of war is more real now, than at any time during the Cold War. Irrationality and instability are rampant worldwide. We must maintain our ability to control the air at any time and place in such a world. *Congress, give us the wings that will stay on, and prevail!*

To guarantee U.S. air superiority the F-22 is one of the most crucial aircraft in the history of the U.S. Air Force. Unfortunately, the F-22 program has been challenged by some members of Congress. Featuring extraordinary maneuverability, stealth, and a super-cruise capability, the F-22 promises to dominate the air well into the next century. We need to get the F-22 operational as soon as possible. Shown are two pre-production prototypes.

Shown here is a mockup of the Boeing design for the Joint Strike Fighter. (JSF). Designated the X-32

The JSF is an important follow-on to the F-22 for air-to-ground operations. The JSF is designed to meet the requirements of the U.S.A.F., U.S.N., the U.S. Marine Corps, and the RAF. The Marine version will have a short takeoff and vertical landing (STOVL) capability.

The Lockheed design of the JSF. Designated the X-35. Note the similarity to the F-22. Much of the technology transfers.

In light of the British experiences in the Falkland Islands War, the author believes the JSF for the U.S. Navy should also use at least some of the STOVL versions, like the U.S. Marine version. British VTOL Harriers were able to operate in the Falklands in conditions that were impossible for conventional carrier aircraft. U.S.A.F. & Raf close support aircraft could also benefit from this STOVL capability.

The Japanese F-2 is a follow-on to the F-16 jointly designed by Lockheed and the Japanese. It's already superior performance is significantly enhanced.

The EF-2000 Eurofighter is superior to the USAF F-15 in many categories. It's low radar signature is a critical factor (See the table on the next page).

Russia, China, and many Islamic Countries are using russian designs for their top fighters. The Sukhoi SU-35 shown here has parity with the F-15 in several key performance aspects (see table on following pages).

The table on the next page dramatically makes the case for the F-22 to maintain U.S. Air superiority. Our top air to air fighter, the F-15, is being equalled in performance all over the world.

Aerial combat without the F-22

If the F-15 Eagle continues as the Air Force's main interceptor, the Air Force claims it will lose much of its advantage over other fighters. Below, a comparison of the F-15 to other fighters it would face in 2000 and 2005. The Fulcrum, Flanker and SU-35 are Russian fighters, the Rafale is French and the EF-2000 is being developed jointly by Germany, the United Kingdom, Italy and Spain.

F-15 advantage ⬆   F-15 parity ⚌   F-15 disadvantage ⬇

In 2000   In 2005

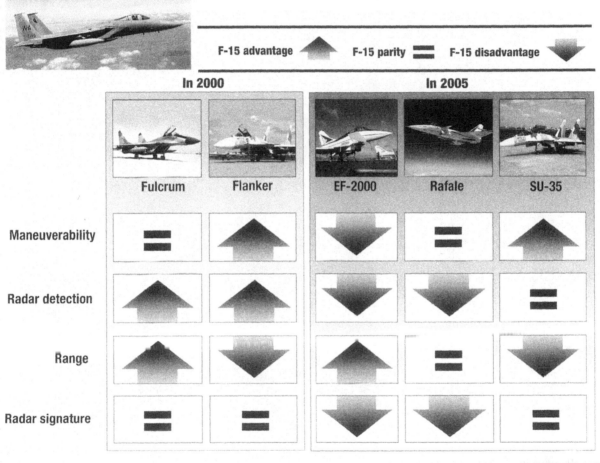

	Fulcrum	Flanker	EF-2000	Rafale	SU-35
Maneuverability	=	⬆	⬇	=	⬆
Radar detection	⬆	⬆	⬇	⬇	=
Range	⬆	⬇	⬆	=	⬇
Radar signature	=	=	⬇	⬇	=

Source: Air Force Times Aug.2, 1999

9 781563 115684